MORE
NUMBER GAMES

By the Same Authors

NUMBER GAMES TO IMPROVE
YOUR CHILD'S ARITHMETIC

More Number Games

Mathematics Made Easy Through Play

Abraham B. Hurwitz
Arthur Goddard
David T. Epstein

Funk & Wagnalls
NEW YORK

To all those who found mathematics a big bore and who prayed for it to disappear magically from the curriculum

Designed by Ingrid Beckman

Manufactured in the United States of America

Library of Congress Cataloging in Publication Data

Hurwitz, Abraham B.
    More number games.

    Includes index.
    1. Mathematical recreations. 2. Mathematics — Study and teaching. I. Goddard, Arthur, joint author. II. Epstein, David T., joint author. III. Title.
QA95.H886   793.7'4   76–3600
ISBN 0–308–10255–X

10 9 8 7 6 5 4 3 2 1

CONTENTS

INTRODUCTION:

To the Parent

The enthusiastic reception accorded our first book, *Number Games to Improve Your Child's Arithmetic,* has encouraged us to produce this sequel. All the games in it are designed to be played by children in elementary school and junior high school.

How This Book Can Help Your Child

Every game in this book has an educational purpose. Each one has been devised to help your child to improve and perfect some mathematical skill. Whether he plays the games with you, with his friends, or by himself, he will not only be having fun but will be learning the basic techniques of arithmetical, algebraic, and geometrical reasoning and computation and the fundamental laws of mathematics.

1

These are subjects that often perplex children and even give rise to feelings of distaste and dread. Gaining skill in mathematics is frequently associated with hours of boredom and tedious drudgery in the performance of dry drills and exercises. Indeed, disagreeable experiences in mathematics classes or with textbooks and homework sometimes result in the formation of a mental block that actually inhibits learning and prevents progress.

Games, on the other hand, can make the learning of mathematics easy and pleasurable. All the games in this book have been carefully selected after being thoroughly tested and found to be effective. Playing them will, in the first place, motivate your child by providing him with an incentive for further and continued learning. They will stimulate his imagination and curiosity about numbers and spatial relations and inspire him to proceed on his own and at his own pace as far as he can go. At the same time, they will improve the accuracy, proficiency, and speed with which he performs mathematical computations. They will spur your child to extend his mathematical abilities. And the challenge of competition will provide additional stimulus to explore this fascinating field even further.

What Your Child Can Learn from These Games

A knowledge of the basic concepts, principles, skills, and laws of mathematics is of evident value to everyone. All of us find mathematical knowledge useful in many everyday activities, such as calculating interest on money in the bank, computing the percentage of a discount in comparison shopping, balancing our accounts, measuring areas and volumes in home decoration and furnishing, reading maps,

2

and planning a budget. We live in a world of quantitative and spatial relations.

In addition, arithmetic, algebra, and geometry find many applications in business, finance, statistics, technology, and the arts and sciences. Indeed, mathematics has rightly been called the queen of the sciences, because virtually every science makes use of quantitative laws and calculations or the application of geometric laws. The development of mathematics in the last centuries has been one of the chief factors in the enormous progress made by the physical and social sciences. Computers have recently opened further vistas: the exploration of outer space and the application of mathematics to psychology, chemistry, sociology, political science, and even music. Without mathematics, the modern world, with its technological marvels, its medical advances, and its labor-saving efficiencies, would be quite impossible.

Thus one of the greatest services you can perform for your child is to give him or her a knowledge and love of mathematics. And this is just what the games included here have been designed to do. Indeed, they may be regarded as an instructional substitute for types of recreation that would otherwise simply kill time.

So, as your child plays a game of dice, dominoes, bingo, lotto, or cards as adapted in this book, or any other game in it, he will at the same time be acquiring one or more of a host of useful mathematical skills. Besides developing proficiency in the basic techniques of arithmetical computation—adding, subtracting, multiplying, and dividing—he will learn to figure percentages; to calculate the interest yielded by a principal sum; to apply the formulas for finding the areas and perimeters of various types of polygons, the areas and circumferences of circles, and the volumes of solids; to use correctly the nomenclature and symbols of geometry; to recognize various spatial figures and relationships; to apply the Pythagorean theorem; to solve different types of algebraic equations; and to perform calculations involving square roots, Fibonacci series, numbers in various base systems, decimals, and fractions. In addition, he will be taught how to perform all these operations with accuracy and speed by means of some shortcuts.

How to Use This Book

Almost all the games in this book were designed to be played in the intimacy of your home, by the child alone, with you, or with members of the immediate family. (A separate chapter is devoted to party games for larger groups.)

No pedagogic skill on your part is necessary. You will not be doing the teaching: *the games will.*

Hence it is wise to keep the accent throughout on fun and recreation and to allow the learning to take care of itself, as it surely will. To this end, we have included a chapter on what we call "mathemagic." You can use this to arouse your child's interest at the outset or to overcome feelings of fear or dislike that he may have developed toward mathematics. With the help of mathemagic you can set the mood of relaxation and entice him into a game. Another chapter is devoted to mathematical oddities, curiosities, "teasers," stunts, and novelties, and still another chapter consists of jokes and riddles. These too can be used to stimulate as well as sustain your child's interest. Some of the tricks and oddities of mathematics also illustrate or emphasize a point, clarify a concept, demonstrate an interesting fact, or aid the child's memory.

Besides, every child—and particularly one who has been experiencing difficulty with mathematics—will welcome the many shortcuts and other useful timesaving devices which have been included in this book. They will provide him with a quick and easy method of performing operations that might otherwise prove tedious and will increase his confidence in his ability to solve mathematical problems.

Accordingly, you should not find it difficult to interest your child in these games. Children like to win a contest. It gives them great satisfaction to outwit an opponent. They rise to a challenge and bend every effort to meet it.

Your teaching role, then, as a parent is primarily to provide motivation and support. You can help your child, at first, to select the games to be played, interest him in them, show him how to play them, play some of them with him, and encourage him with your help and praise.

Selecting the Games

The order in which these games are to be played depends on the level of ability at which you find your child, his rate of progress, and his interests. You guide him by selecting a game that will interest him at the moment, teach him the principle or skill that he needs to learn, and stimulate him to proceed to other games or variations at ever higher levels of aspiration and difficulty.

In planning his "curriculum," you will want to take into consideration the following factors for each game:

1. skills taught
2. level of maturity
3. materials needed
4. number of players
5. method of play

For every game information about each of these factors is readily available.

Skills Taught

The skills taught by each game are listed under its title. For example, under the title The Knight Game you will find these skills listed: counting, spatial relations, addition of fractions, construction of a Fibonacci series.

At the back of the book you will find an Index of Skills Taught that enables you to locate at a glance just the game you need to correct your child's weaknesses in mathematics, to reinforce previous learning, or to extend his mathematical abilities into new domains, such as solving algebraic equations or calculating the areas of polygons.

Incidentally, throughout the book we have used the expression "MADS" as a convenient acronym formed from the four fundamental operations of arithmetic — multiplication, addition, division, and subtraction. These basic skills are, of course, reviewed and reinforced in the games in this book.

Level of Maturity

The games in this volume are designed to be played by children in elementary school and junior high school.

It is difficult if not impossible to determine the precise age for which any of these games is ideally suitable. In the first place, children of the same age are not always equally mature. They differ in experience, schooling, rate of intellectual growth, and mathematical abilities and interests.

Besides, as you will discover, most of these games can be profitably and enjoyably played at more than one level of maturity. The same game that appears challenging or difficult to a twelve-year-old can prove easy for a child of fourteen and may be intriguing even for adults, especially if the level of difficulty is "escalated" by means of the successive variations described in the text. You will find that many of these games, especially those in the chapter entitled "Party Games," are ideal for families and large groups. Here the spread in ages and mathematical skills is actually an advantage, because older and younger children can enjoy playing together, and you can join in the fun too.

You will soon learn from experience at just what level of maturity and mathematical ability your child is to be placed, and you will have the pleasure of observing him progress from one level to the next as he acquires proficiency through experience in playing the games.

Within each chapter the games have been arranged in order of increasing difficulty. A game should be neither so difficult as to frustrate a child's best efforts to rise to its challenge

7

nor so easy as to appear uninteresting. You will usually be able to adjust the difficulty of a particular game to the individual capacities of your child. You can make it easier by simplifying the rules or using smaller numbers as examples; and you can make a game harder by playing the variations we have added to many, which offer a means of graduating the difficulty of the intellectual challenge. It is advisable to begin by playing the first game in a series before attempting any of its variations. You can then simplify or complicate the game as you desire.

Materials Needed

Most of the games have been devised so that a minimum of preparation is required by parent or child.

Chapter 1, as its title indicates, consists entirely of paper-and-pencil games. Where graph paper is needed, it is mentioned under the title of the game.

Ordinary playing cards are the only materials needed for playing the games in chapter 2.

Dice or dominoes are needed to play the games in chapter 3.

The games that require you to prepare materials are for the most part to be found in chapter 4, though a few may be found in chapter 5, as specified under the name of each game.

Naturally you will want to have these materials ready in advance. Preparing them is easy.

Suppose, for example, that the materials called for are cards prepared with numbers, symbols, problems, geometrical figures, or algebraic equations. You can produce these simply by taking an old deck of playing cards and pasting on the front of each card a detachable sticker on which the necessary notation can be written in black ink with a felt-tipped marking pen. Self-adhesive file folder labels are ideal for this purpose.

Another method of preparation is to use 3-by-5-inch cards cut up into 1-inch squares, on each of which a number, symbol, problem, or figure can be placed. From fifteen 3-by-5 cards you can make 225 squares. In the directions you will find exact information about the number of cards required and the notation to place on each card. The task of preparing the cards can be made into a learning activity in itself if you encourage your child to print the numbers and signs and draw the figures or diagrams needed.

For some of the games, where a time limit has been suggested for each answer in order to encourage quick responses, the word "timer" has been included among the materials listed under the title. You may use a wristwatch, stopwatch, clock, disk bell, hourglass egg timer, gong, or electric cooking timer.

Number of Players

Some of these games are for one player. Others are for two, three, or more, or are best played by large groups.

You can see at a glance how many players each game is suited for by looking under its title.

Method of Play

For each game and the variations, if any, that follow it the method of play is described step by step.

Before introducing a game to your child, read the description of it yourself and become familiar with the possible variations.

Next prepare your materials if any are needed.

Then name the game and explain slowly and clearly as much of it as is needed to begin playing. It is wise to keep the rules brief at first. You can always add other rules and complications after your child has mastered the basic pattern.

Since in several of these games the first player, as in chess, enjoys an advantage, the players should take turns going first.

It is a good idea to join in the game yourself, at least at the beginning. Your participation and enthusiasm can hardly fail to prove infectious. Later, if the child prefers to play by himself or with his peers, you may reduce your role to that of umpire or scorekeeper, or you may delegate the responsibility to an older child under your supervision.

If the child's efforts are successful, praise him. If he tends to become discouraged, remind him of his past successes. If he fails to solve a problem, give him the answer, show him how you calculated it, and continue with the game. If you always correct mistakes promptly and kindly and answer questions simply, clearly, and patiently, your child

10

will learn rapidly, and the two of you will build a close, warm relationship conducive to continued progress. For you, there will be the experience of sharing the fun of learning, of joining in the games, of spending many delightful hours over the years in close companionship with your child. For the youngster, there will be the eager anticipation of playing new games or variations with you, and in later years your child's memories of you will be associated with the happy times spent in your company.

Other Teaching Aids

For your convenience, you will find in the Appendix a list of symbols commonly used in geometry, formulas for calculating the areas of common geometric figures and the volumes of common geometric solids, a table of metric equivalents, a table of squares and square roots, and the values of the trigonometric functions. At the end there are an Index of Skills Taught and an Index of Games.

The symbols, formulas, and tables can be useful in preparing materials for many of the games and can be referred to in playing them. As our country is in the process of changing to the metric system, familiarity with its features and with the metric equivalents of our traditional measures of length, volume, and weight will become indispensable in all the computations of daily life.

The alphabetical Index of Skills Taught will enable you to find just the game you need to correct your child's weaknesses or to build on the knowledge and skills he has already acquired.

The alphabetical Index of Games will help you to locate games in the book which you have enjoyed playing and whose description you wish to find again.

PART ONE

Games To Play

1

Paper-and-Pencil Games

EVEN IN THIS AGE of desk-top and pocket calculators, the average person still finds it necessary to use paper and pencil to perform many mathematical computations and to record their results. Paper and pencil are certainly indispensable in the solution of problems of geometric construction and the analysis of spatial relations and coordinates, charts, and graphs.

The games in this chapter will accustom the players to employing written figures, signs, and the conventional symbols — arithmetical, algebraic, and geometric — needed in mathematics and will help develop speed and agility in their use.

Figure Patterns

Addition, Subtraction

Two players

Paper and pencil

Each player makes his own ticktacktoe pattern and inserts the digits 1 through 9 in random order, using each only once.

Now the players exchange papers.

Drawing connecting lines as shown, each player then links possible combinations of two or more adjacent boxes that add up to 15. The same box can be included in more than one combination.

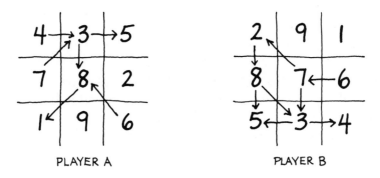

PLAYER A   PLAYER B

In A's pattern, as illustrated, B finds three combinations: $7 + 3 + 5 = 6 + 8 + 1 = 4 + 3 + 8 = 15$. Using B's pattern, A finds four combinations: $6 + 7 + 2 = 7 + 3 + 5 = 2 + 8 + 5 = 8 + 3 + 4 = 15$.

16

Thus B scores 3 points, and A scores 4. If either player notes a combination not seen by his opponent (such as 7 + 8 in A's pattern or 9 + 6 in B's pattern), the opponent loses 2 points.

Several rounds are played until one player scores a total of 25 points.

The game can be made more difficult by using larger numbers—for example, 11 through 19—and raising correspondingly the sum to be reached (for example, 30 instead of 15).

Box Score

Addition, Subtraction, Multiplication, Division, Use of Coordinates, Squares, Square Roots

Two players   Paper and pencil

First, each player prepares for himself a square with three boxes on a side.

Into the nine boxes thus created he inserts the numbers 1 through 9 at random, without showing his square to his opponent.

Finally, each player indicates the horizontal and vertical directions from the lower left-hand corner of his square by writing the numbers 1, 2, and 3, first from bottom to top, and then from left to right, outside the square, as shown in the illustration. These numbers are coordinates determining the location of any box. For example, the combination (1, 1) signifies the first box (the one farthest to the left) in the bottom row, and the combination (3, 2) denotes the last box (the one farthest to the right) in the middle row in each square.

The first player now calls out two sets of coordinates — say, (1, 2) and (3, 3), designating boxes in his opponent's square. The latter must then state what numbers in his square are in the boxes located by these coordinates. In the illustration, they are 2 and 7. So the first player scores 9, the sum of 2 and 7.

Now the second player follows the same procedure. If he asks his opponent for (3, 2) and (1, 2), the second player will get 9 and 1, for a total of 10.

No player may call the same set of coordinates twice. The first player, when his turn comes again, can call (1, 1), but not (1, 2) or (3, 3).

The game continues for four rounds; the player with the higher total wins.

There are a number of ways in which the game can be increased in difficulty:

1. Each plays calls out *three* sets of coordinates in each of three rounds.

2. Each player *subtracts* the smaller from the larger of the two numbers he gets from his opponent.

3. Each player *multiplies* the two (or three) numbers he locates on his opponent's square.

4. Each player *divides* the larger by the smaller of the two numbers he gets from his opponent, rounding off to the higher whole number if necessary. Thus, if a player gets 6 and 2, he earns 3, while if he gets 9 and 2, he earns 5.

5. The players can insert squares and square roots in the boxes. For example, 2^2, 3^2, $\sqrt{4}$, and $\sqrt{9}$ may be used instead of 4, 9, 2, and 3. The player who locates any of these numbers has to make the correct calculation in order to earn the point value.

6. Best of all, the number of boxes can be gradually increased. In the next illustration each player has prepared a square consisting of four boxes on each side (sixteen boxes in all) and has inserted the numbers from 1 through 16 in them at random, sometimes expressing them as squares or square roots: $4 = 4^1$ or 2^2 or $\sqrt{16}$. At first, the players can call for just two numbers to be added or multiplied, or the smaller to be subtracted from the larger, or the

19

larger to be divided by the smaller, and the match ends after eight rounds. In the next match they can call for four numbers, to be added or multiplied, and play four rounds. In this way, all basic arithmetical operations can be practiced with squares of varying size.

PLAYER 1

4	$\sqrt{9}$	12	13	1
3	$\sqrt{49}$	11	2	15
2	4^2	5	$\sqrt{36}$	14
1	$\sqrt{64}$	$\sqrt{100}$	9	4
	1	2	3	4

PLAYER 2

4	16	7	6	13
3	$\sqrt{25}$	10	$\sqrt{81}$	11
2	3	$\sqrt{16}$	12	$\sqrt{1}$
1	8	$\sqrt{4}$	15	14
	1	2	3	4

The Knight Game

Counting, Spatial Relations, Addition of Fractions, Construction of a Fibonacci Series

Two or more players   Timer

Paper and pencil

The knight is a chess piece that moves two boxes in any direction and one box at right angles to it.

20

All the players use the same twenty-five-box square.

The game begins when the first player inserts the number 2 in any box in the square. Now his opponent, counting by twos (2, 4, 6, 8, . . .), must place the next number in the series (4), within a given time limit, in any box to which a knight can move from the box containing the number 2. Only one number can be placed in a box.

In the games illustrated, there are only two players, the second of whom, to distinguish his numbers from those of his opponent, circles them. If more than two players participate, each is given a differently colored pencil to use in writing his numbers in the boxes.

Sooner or later a player is stalemated: he cannot make a move because all the spaces to which a knight can move are already occupied by numbers.

The player who forces the stalemate wins the round. He scores the value of the highest number reached. In the first game illustrated, the first player scores 24 points; in the second game, the second player scores 38 points.

	2	→		(24)
		14	(4)	
(12)	6	←	22	(16)
		18	(8)	
	10			(20)

	38	(4)	18	
2	(20)	14	(36)	6
(12)		(28)	22	(16)
30	(24)	38	(8)	34
	10	(32)	26	

If a player cannot find an available box within the time limit, places his number in a box a knight cannot reach, or inserts the wrong number, he is penalized the value of the number he has reached or should have reached.

In the second round the players count by threes: 3, 6, 9, 12, ... The game can be escalated gradually in this way as desired.

Another way of complicating the game is to have the players count in geometric progression, multiplying each number by 2 to get the next term in the sequence: 1, 2, 4, 8, 16, 32, . . . ; or by 3: 1, 3, 9, 27, ... Or they can increase the number of boxes in the square. Two games with thirty-six-box squares are illustrated, the first counting by twos, the second, by threes. Squares with a larger number of boxes can also use geometric series.

		6	(12)	18	
(40)	26	(16)	30	(8)	
(24)		(4)	10	(20)	14
38		22	(28)		(32)
	2		34		
	(36)				

(6)		(12)	27	(72)	51
15	(24)	3	(48)	33	
	9	(30)	21	(54)	69
	(18)	63		45	(36)
		(42)		(66)	57
			(60)	39	

The most challenging form of this game involves counting by constructing a Fibonacci series of fractions, in which each new term is the sum of the preceding two terms. After the first player inserts *two* fractions in any box, the second player inserts their sum, the third fraction in the series, in any box to which a knight can move from the first box.

If the first fractions are 1/10 and 1/5, the second player must add the two and place their sum in an appropriate box. A quick way to perform this addition — and indeed, to add any two fractions — is to multiply the denominator of the first fraction (10) by the numerator of the second (1), and the denominator of the second fraction (5) by the numerator of the first (1). Add the products: $(10 \times 1) + (5 \times 1) = 10 + 5 = 15$, and divide the sum by the product of the denominators: $\dfrac{15}{10 \times 5} = 15/50 = 3/10$. Or change to the lowest common denominator (tenths) and reduce the final answer, if necessary: $\dfrac{1}{10} + \dfrac{1}{5} = \dfrac{1}{10} + \dfrac{2}{10} = \dfrac{3}{10}$.

Player 1 has to calculate the next term after 3/10 in this series: $1/5 + 3/10 = \dfrac{(5 \times 3) + (10 \times 1)}{5 \times 10} = \dfrac{15 + 10}{50} = 25/50 = 1/2.$

Player 2 continues with $3/10 + 1/2 = \dfrac{(10 \times 1) + (2 \times 3)}{10 \times 2} = \dfrac{10 + 6}{20} = 16/20 = 4/5$; then Player 1 must add 1/2 and 4/5.

Excitement can be added by shortening the time limit for each successive round.

Each round should begin with a different pair of fractions. For example, the second round might start with 1/8 and 1/4; then the third term would be 3/8; the fourth, 5/8; and the fifth, 1.

23

If the third round begins with 1/12 and 1/6, the series continues with 1/4, 5/12, 2/3, 1 1/12, 1 3/4, and 2 5/6.

The great Swiss mathematician Leonhard Euler (1707–83) composed a complete knight's tour of the chessboard, writing the numbers from 1 through 64 in the boxes of a square with eight boxes on a side:

1	48	31	50	33	16	63	18
30	51	46	3	62	19	14	35
47	2	49	32	15	34	17	64
52	29	4	45	20	61	36	13
5	44	25	56	9	40	21	60
28	53	8	41	24	57	12	37
43	6	55	26	39	10	59	22
54	27	42	7	58	23	38	11

Many other solutions are possible.

24

Magic Squares

Addition, Subtraction, Multiplication, Division, Fractions, Decimals, Roman Numerals

Two or more players   Cards

Pencil   Timer

A magic square is a square divided into boxes containing numbers whose sum in any row, column, or diagonal is the same.

Each player begins by preparing a card with a nine-box magic square. This is easy. Just write a number — say, 9 — in the center box. Multiply the number by 3 (the number of boxes on each side of the square). The product is 27. Now find any three numbers, not including 9, that add up to 27 — say, 5, 10, and 12. Write them in the first column, as shown.

Now it is simple to add new numbers, not already used, to make the magic square. For example, $5 + 9 + x = 10 +$

$9 + y = 12 + 9 + z = 27$. Enter x (13), y (8), and z (6) in the first diagram. Now you can make and solve two more equations: $5 + 6 + n = 12 + 13 + m = 27$. With n (16) and m (2) inserted in the appropriate boxes, the complete magic square looks like this:

5	16	6
10	9	8
12	2	13

You can make a second magic square with the same number of boxes, but with larger numerals, by adding the same number to the number in each box. Suppose you add 4 to every number in the first square; the result would be a square in which each column, row, and diagonal adds up to 39:

9	20	10
14	13	12
16	6	17

Similarly, if you multiply the number in each box of a magic square by the same number, you produce a new magic square with larger numbers. Try multiplying every number in the first magic square by 2. Each column, row, and diagonal will add up to $27 \times 2 = 54$:

26

10	32	12
20	18	16
24	4	26

Now try dividing each number in the first magic square by 2. Now the new magic square will contain some fractions. The magic constant (the sum of each column, row, and diagonal) is 27/2:

$2\frac{1}{2}$	8	3
5	$4\frac{1}{2}$	4
6	1	$6\frac{1}{2}$

Subtracting the same number from every number in the first magic square works too. If you subtract 6 you will get some negative numbers:

-1	10	0
4	3	2
6	-4	7

The magic constant is 9.

After each player has made up his own nine-box magic square and written it on his card, he copies it on the back of the card, but omits the numbers in five boxes, as illustrated. The numbers in all the boxes of one column, row, or

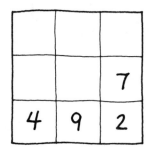

diagonal should be given, plus the number in one other box. If there are more than two players, each player should write his name or initials on the reverse of his card.

The cards are exchanged, with the reverse (or challenge) side up, so that no player has his own card. The players then set about filling in the missing numbers. (Of course, no one is allowed to turn over his card until time is called.)

1 point is scored for each correct number found within the time limit.

At a higher level, the challenge side of the card may contain numbers in only three boxes (a complete row or column):

The sum of the three numbers given in the challenge should be divisible by 3: $15 + 7 + 14 = 36$. To meet this challenge, divide the magic constant by 3: $36 \div 3 = 12$. Enter 12 in the central box of the square. Now it is easy to calculate and enter the remaining numbers.

Here are some examples of challenges at various levels of difficulty:

8	4.5	
5.5		
6		

14		
	11	
10		8

X	III	
	VII	
		IV

		3
	2½	
2		1

	2	
	10	14
	18	

		11
		12
		7

The high-scoring player wins the round. Several rounds should be played with magic squares containing larger numbers, decimals, fractions, Roman numerals, etc.

Algemagic

Algebraic Equations, Square Roots

Two or more players   Cards

Pencil   Timer

In this variation of Magic Squares, the numbers in the boxes are represented, on the challenge side of each card, by the letter *x* in algebraic equations. The other rules are those of Magic Squares.

Here is an example, giving both sides of the card.

10	15	14
17	13	9
12	11	16

$\frac{3x}{5} = 6$	$\frac{2x}{3} = 10$	$(3 \times 4) + \sqrt{4} = x$
		$x + 5 = 14$

Equimagic

Finding Roots, Powers, Fractions, Solving Algebraic Equations

Two, three, or four players   Disks or buttons

Paper and pencil   Cards

You can begin with the familiar nine-box magic square illustrated. It is placed on the table for all the players to see.

8	1	6
3	5	7
4	9	2

Each player receives as many cards as there are boxes in the magic square—in this case, nine. For every number in the magic square, each player writes on a card a mathematical expression equivalent to it. One player may represent 8 by 2^3; another by $\sqrt{64}$. Moving along the top row, 1 may be written as 9^0 or as $9/x = 9$, and 6 as $18/3$ or as $3^2 - 3^1$. In the second row, 3 may be shown as $x/3 = 1$ or as $2x/3 = 2$; 5, as 5/8 of 8 or as $x - 3 = 2$; 7, as $7x = 49$ or as $5x + 2x = 49$. Finally, equivalents of 4 include $3x - 2 = 10$ and $5^2 - 21 = x$; 9 may be denoted by $4^2 - 7$ or by $\sqrt{81}$; and 2, by $\sqrt{4}$ or $3x + 2x = 4x + 2$.

When all the cards are prepared, they are thoroughly shuffled and placed in a pile face down on the table.

The first player now draws one card. If it is a card he prepared, he discards it at the bottom of the pile and draws again. If it is not, he solves the problem on it and places a disk or a button on the box in the magic square containing the number equivalent to the solution. For example, if he draws a card with $\sqrt{64}$ on it, he calculates $\sqrt{64} = 8$, and then covers the box containing the 8. The round continues until a player completes a column, a row, or a diagonal. (The other disks on the line need not be his.)

31

Each player scores the sum of the numbers on which he has placed disks. The player who completes a line scores the number he has just covered plus, as a bonus, the value of all the numbers on the line. Thus if a player completes the diagonal 8, 5, and 2 by covering 5, he scores $5 + (8 + 5 + 2) = 20$.

For each round use a different magic square. The first player to score 50 points wins.

2

Card Games

THERE IS NOTHING like a game of cards to capture and hold the interest of children. They enjoy playing such familiar games as Rummy, Casino, and Twenty-one, and they quickly become accustomed to sorting the numbers on the cards into sets, arranging them in sequences, and adding them.

The games in this chapter turn the pleasures of cardplaying to educational advantage. They review the basic operations of arithmetic; teach the configurations and elementary formulas of geometry; and provide practice in the solution of algebraic equations.

Target Practice

Addition

Two players　　　　　　Two decks of playing cards

Each player shuffles a deck of cards and holds it face down in his hand.

Any number from 13 through 50 — say, 29 — is agreed on as the "target." (A king counts as 13; a queen, 12; a jack, 11.)

The first player aims for the target by turning up two or more cards from the top of his deck, one at a time, and laying them in a column on the table, as shown, looking for a combination of successive cards that add up to 29. Player A places the four cards illustrated $(3 + 7 + K + 6 = 29)$

```
        Player A          Player B
          10                 7⌉
          Q                  Q⎬   29
          3⌉                10⌋
          7⎪                 5
          K⎬     29          2
          6⌋
```

in his scoring pile, leaving the rest on the table. He continues to lay down cards with the same object until he has used up all the cards in his hand. If he fails to notice that a combination of successive cards in his column adds up to the target number and leaves them on the table, laying down one more card, his opponent can point this out and take the cards for his scoring pile.

The second player follows the same procedure. He places the first three cards he has laid down in a separate pile: $7 + Q + 10 = 29$.

When the players have dealt out their decks, each scores a number of points equal to the number of cards in his scoring pile.

A new round begins with another number as target, and with each player beginning with a full deck. The winner is the player with the higher score after five rounds.

Bull's-Eye

Subtraction

Two players   Two decks of playing cards

In this variation of Target Practice, the bull's-eye — any number from 1 through 5 — must be hit by subtraction.

In the illustration, with 4 as the bull's-eye, the first player was able to pick up $7 - 3 = 9 - 5 = 9 - (8 - 3) = 4$, while the second player collected $5 - 1 = J - 7 = K - 6 - 3 = 4$.

Player A		Player B		
8		J		
7 }	4	5 }	4	} 4
3		1		
9 }	4	7		
5		K }		
9 }	4	6 }	4	
8		3		
3				

Note that after Player B picked up the 5 and the ace, the 7 he dealt became the next card after the jack and could be picked up with it.

Straight Shooting

Multiplication

Two players   Two decks of playing cards

A straight shooter in this variation of Target Practice has to hit, by multiplication, any target taken from the multiplication table—for example, 10, 12, 16, 24, 30, 32, 40, 48, 54, 60, 64, 72, 80, 90, 100.

If the target is 48, a player can pick up $Q \times 4$ or $6 \times 4 \times 2 = 48$.

Sharpshooter

Division, Fractions

Two players   Two decks of playing cards

A player has to be sharp to hit the mark in this variation of Target Practice. Targets are from 1 through 5. Shooting is by division, rounded off to the nearest whole number. Otherwise, the game is played exactly like Target Practice.

If the target is 3, the following combinations can be picked up: $6 \div 2 = 3$; K or $13 \div 4 = 3\ 1/4 = 3$; $10 \div 3 = 3\ 1/3 = 3$.

Potshots

Addition, Subtraction, Multiplication, Division

Two players   Two decks of playing cards

This is our "MADS" variation of Target Practice — that is, the players may use one, two, three, or four operations in any combination of multiplication, addition, division, and subtraction, to hit the target, which can be any number from 1 through 50.

If the target is 25, a player can pick up all sorts of combinations, such as $(6 \times 4) + 1 = (9 \times 3) - 2 = K + 6 + 8 - (4 \div 2) = 25$.

Points for Points

Decimal Operations

Two players   Timer

Deck of playing cards

In this game the players earn points for decimal points correctly placed after quick calculation.

The deck should be divided into two packs, one consisting of the black suits (clubs and spades) and the other of the red suits (hearts and diamonds). Both packs should be thoroughly shuffled and placed side by side face down on the table.

The first player draws two cards, one from the top of each deck. The numbers on the black cards are considered whole numbers; those on the red cards are decimal numbers. So if the first player draws black 3, red queen (equivalent here to 0.12), he has 3 and 0.12. A jack is 11 or 0.11, depending on its color; a king is 13 or 0.13.

Within a given time the player must multiply his whole number and his decimal correctly to attain his score. The first player's score is $3 \times 0.12 = 0.36$. If a players fails to calculate the correct answer, he scores nothing.

The players take turns drawing cards and multiplying whole numbers by decimals until all the cards are drawn. The winner is the player with the higher total score, rounded to one decimal place. The time limit can be made shorter after each round or even after each turn as the players gain facility in decimal calculation.

A variation of the game requires each player to subtract his decimal from his whole number. Thus, if he draws a 4 and a 0.9, his score is 3.1.

In another variation the whole numbers are divided by the decimal numbers. Thus, if a player draws black 8, red 0.7, he divides 8 by 0.7. The quotient, 11.43, is rounded off to one decimal place, 11.4, which is his score if he performs the calculation in time.

Collections

Addition, Subtraction, Multiplication, Division

Two, three, or four players   Timer

Deck of playing cards

After the deck has been shuffled, the cards are divided into as many equal packs as there are players. (Any leftover cards are set aside.) Each player places his pack face down in front of him.

The jack again counts as 11, the queen as 12, and the king as 13.

Each player draws one card from the top of his pack and places it face up in the center of the table. If there are two players, the cards turned up may be the 5 of hearts and the 6 of spades.

The first player now turns up the two top cards in his pack. Let us say that they are the queen (12) of hearts and the 2 of clubs. With the aid of "MADS" (multiplication, addition, division, and subtraction), he must try to make the two cards he has turned up equal to one of the two cards exposed on the table. He has four results to choose from: $12 \times 2 = 24$; $12 + 2 = 14$; $12 \div 2 = 6$; and $12 - 2 = 10$. Only division results in a number equal to the value of one of the cards on the table, the 6 of spades. If the player points this out within the given time limit, he collects the three cards — the two he drew and the card on the table whose value they equal — and places them in a pile to one side. He turns up one card from the top of his pack and puts it on the table face up to replace the one he collected.

The next player exposes two cards from his pack and continues. If a player cannot collect any cards, he places those he has turned up, face down, under his pack of still unused cards, for later use.

The game proceeds until one player has collected or discarded all the cards in his pack. The players add the value of their collected cards, and the player with the most cards collected adds 10 to his score. High score wins.

Battle of Equals

Equivalents, Addition, Subtraction, Multiplication, Division

Two, three, or four players

Adhesive labels (52 per player)

Deck of playing cards

Write on each removable adhesive label or tape a statement equivalent to the numerical value of the card to which it will be attached. Thus 4 may be represented by 16/4, 8/2, $18 - 14$, $1 + 3$, or any similar expression. Four expressions will be needed for the 4's, and four for each other numerical value.

Attach the labels to the appropriate cards; then divide the deck evenly among the players, dealing one card at a time. (If there are three players, discard the leftover card.) Each player stacks his cards face down in front of him.

The first player draws a card from the top of his stack, reads the statement attached to its back, calculates the equivalent number, states it, and turns the card face up on the table as a check on his arithmetic. The next player does the same. The one whose card has the higher numerical value takes both cards and places them face down at the bottom of his stack. If the cards match in value—let us say, both are 6—they are left on the table face up. Each player puts another card over the first one, and the player with the higher card takes in all four cards. If the second two cards are equal, the battle continues until one player lays down a card with a higher numerical value than his opponent and takes all the cards lying face up.

When one player has lost all the cards in his stack, the other wins.

If a player errs in calculating the numerical value of the statement on the back of a card, he must surrender the card to his opponent. If more than two play, the next player now battles the winner, until one player wins all the cards.

Equivalences can also be expressed as "the number of

pints in a quart" — or quarts in a gallon, feet in a yard, millimeters in a centimeter, items in a dozen, sides in a hexagon, angles in a triangle, equal sides in a square, weeks in a month, days in a week, etc. Thus, a Time, Weight, and Space Battle can be played as a variation of the Battle of Equals.

Exact Reckoning

Addition, Subtraction, Multiplication, Division

Two, three, or four players   Timer

Deck of playing cards

In this game a jack has the numerical value of 11; a queen, 12; and a king, 13.

After shuffling the cards, place the deck on the table face down.

The first player turns up five cards from the top of the deck. Within a given time limit he is to carry out whatever arithmetical operations — addition, subtraction, multiplication, or division — are necessary to make a true mathematical statement using the numerical values of all five cards.

Suppose he draws 9, 8, queen (12), 4, and 7. With these

values he can make the following true statement: $12 - 8 = (9 + 7) \div 4$.

The second player may turn up a king (13), 7, 6, 3, and 4. He can make this true statement: $13 - 7 = 3 \times (6 - 4)$.

For each true statement a player scores 1 point. If he cannot make a true statement within the allotted time, or if he makes a false statement, he scores 0, and the next player tries, within the same time limit, to use the cards surrendered by the first player to produce a correct mathematical equation. If he is successful, he score 2 points. But if he does not choose to make use of the cards discarded by the previous player, he can draw his own five cards from the deck.

The first player to score 10 points wins.

To increase the difficulty you can reduce the time limit with each successive round and gradually add to the number of cards drawn by each player: six, seven, etc.

Factor Sum

Addition, Multiplication, Division

Two, three, or four players   Timer

Deck of playing cards

Again, the jack counts 11; the queen, 12; and the king, 13.

Shuffle the cards and place the deck on the table face down within easy reach of the players.

Each player draws two cards. He places them face up and, within a given time limit, must multiply their numerical values, determine all the factors of the product — that is, all the numbers, except 1 and the number itself, that it can be divided by with no remainder — and add these factors to determine his score for the round.

If the first player draws an 8 and a 5, the product is 40. The factors of 40 are 2, 4, 5, 8, 10, and 20. (Notice how they make pairs: 2 and 20, 4 and 10, 5 and 8.) The sum of the six factors is 49, which is the first player's score.

If the second player draws a 9 and a 4, his product is 36, which is divisible by the factors 2, 3, 4, 6, 9, 12, and 18. Their sum is 54.

The game continues until all the cards are drawn. High score wins.

The Bids Are "MADS"

Addition, Subtraction, Multiplication, Division

Two or more players   Timer

Deck of playing cards

"MADS" — multiplication, addition, division, and subtraction — are used in this game, singly or in any combination, to arrive at the number bid.

As before, the jack scores 11; the queen, 12; and the king, 13.

Shuffle the cards and place the deck face down on the table.

The first player draws four cards from the top of the deck and keeps them face down until he bids. He can bid from 1 through 9.

Let us say he bids 3. Now he turns his cards over. Within a given time limit he must, with the aid of one or more of the fundamental operations of arithmetic, use the four cards to produce a statement equaling 3.

If the cards he turns over are king, 10, 8, and 4, he can produce the equation $\dfrac{(13 - 10) + 8}{4} = 11/4 = 2\ 3/4 = 3$ (rounded off to the nearest whole number).

For his success he scores 5 points. If, within the time allotted, he cannot arrive at the number he bids, he scores nothing, and the next player can attempt a solution of the first player's problem. If successful, the next player scores double, that is, 10 points. Or he can choose his own four cards and follow the first player's procedure. If he bids 5 and gets a hand with 7, queen, 3, and jack, he can solve his problem with the following equation: $(12 + 7) - 3 - 11 = 19 - 3 - 11 = 5$.

The round continues until one player has scored 25 points. To offset any advantage that a player may have from being first, the players can take turns in beginning each successive round until all have had a chance to be Player 1. High score wins the game.

Square Deal

Addition, Subtraction, Multiplication, Division

Two, three, or four players   Timer

Deck of playing cards

After shuffling the cards, the dealer places nine of them face up on the table in a three-by-three square.

The first player, using the three cards of any row, column, or diagonal, tries to perform any two arithmetical operations on their numerical values—addition, subtraction, multiplication, or division—to make them equal to 13 (rounding to the nearest integer if necessary). If he can do this, he scores 5 points and removes the cards. If he can use any three remaining cards to add up to 13 with two operations, he scores 2 points. Of the three cards left, he may use any number, with any number of operations, to reach 13; if he does, he scores 1 point. A time limit should be set for the solution of these problems.

Each player follows the same procedure with his Square

Deal. Several rounds may be necessary. The first player to score 20 points wins.

Suppose, for example, that a player's Square Deal looks like this:

$$9 \quad 3 \quad 7$$
$$8 \quad 2 \quad K\,(13)$$
$$11 \quad 5 \quad K\,(13)$$

First, he removes the top row: $9 + 7 - 3 = 13$. 5

Second, he removes the two kings and the 2: $(13 + 13) \div 2 = 13$. 2

Finally, he removes the 8 and the 5: $8 + 5 = 13$. $\underline{1}$

Total score: 8

Another player may have this layout:

$$4 \quad K \quad 3$$
$$J\,(11) \quad 5 \quad Q\,(12)$$
$$8 \quad 10 \quad 1$$

First, he removes the leftmost column: $(8 \div 4) + 11 = 13$. 5

Second, he removes the middle vertical column: $13/5 + 10 = 12\ 3/5$, which rounded off $= 13$. 2

Finally, he removes the queen and the 1: $12 + 1 = 13$. $\underline{1}$

Total score: 8

The same game can be played with a larger target number — say, 15:

$$7 \quad 9 \quad Q$$
$$1 \quad 9 \quad 8$$
$$3 \quad K \quad 4$$

47

The player scores 5 points by removing the diagonal formed by the 3, 9, and queen: $(9 \div 3) + 12 = 15$.

He scores 2 points by removing the remaining 9, the 7, and the 1: $9 + 7 - 1 = 15$.

Finally, he scores 1 more point by removing the king, 8, and 4: $13 + (8 \div 4) = 15$.

His total score is also 8 points.

Another way to escalate the difficulty of the game is to increase the size of the Square Deal from nine cards to sixteen or twenty-five. Of course, the target number must be raised, too.

Quotients

Addition, Division, Fractions, Decimals, Factoring

Two, three, or four players   Timer

Deck of playing cards

Remove all the 10's, jacks, queens, and kings, so that only the numbers from 1 through 9 are left.

Shuffle the cards and place the deck face down on the table.

Each player, in turn, draws two cards. He treats the numbers as digits in a two-digit number, arranged either way: if he draws a 5 and a 1, he can treat them as 15 or 51. If he chooses 15, he must, within a given time, determine its factors (other than 1 or 15)—in this case, 3 and 5—and add them, scoring 8. But if he had chosen 51, he could have scored 20, the sum of 3 and 17, the divisors of 51. So he should pick his two-digit arrangement carefully in order to score the most points.

A variation of the game forms fractions with the smaller of the two numbers drawn as the numerator and the larger as the denominator. The fractions, converted to decimal form by division, are the players' scores.

If two are playing, 200 points wins; if three, 150 points; if four, 100 points. If there is no winner after all the cards have been turned up, reshuffle the deck and continue until a player wins.

Sets and Series

Addition, Subtraction, Multiplication, Recognition of Sets and Series

Two or more players   Timer

Deck of playing cards

In this game, a *set* is a collection of two or more repeated numbers, such as 3, 3, 3. A *series* is an arrangement of three or more numbers having either a common difference (5, 7, 9, or 4, 7, 10) or a common ratio (2, 4, 8, or 1, 3, 9). The first type of series is arithmetic; the second, geometric.

Shuffle the cards and deal nine cards, one at a time, to each player.

Within a given time limit, each player produces as many sets and sequences with his cards as he can. The same card may be used in several sets and sequences.

Suppose a player is dealt 1, 5, 9, 12, 8, 3, 9, 2, 10. (The jack is equivalent to 11; the queen, to 12; the king, to 13.)

He adds the values of all the cards in all the sets and series he can construct. Set: $9 + 9 = 18$. Arithmetic series: $1 + 2 + 3 = 6$; $1 + 3 + 5 = 9$; $8 + 9 + 10 = 27$; $8 + 10 + 12 = 30$. (Do you see the series the player missed?) Geometric series: $1 + 3 + 9 = 13$. Adding the totals: $18 + 6 + 9 + 27 + 30 + 13 = 103$. In this case, no card was left unused. If cards are not used in a set or series, their sum is subtracted from the player's score. High score wins the round. A bonus of 10 points is scored if the cards in a set or series are of the same color, and 20 points if they are of the same suit.

In later rounds the number of cards dealt can be reduced to eight or seven.

The player with the higher or highest score after ten rounds wins the match.

It is interesting to note that card games like Gin Rummy use sets and series, but the only series used are arithmetic series with common differences of 1.

50

"MADS" Sets and Series

Addition, Subtraction, Division, Multiplication, Recognition of Sets and Series

Two or more players   Timer

Deck of playing cards

Shuffle the cards and deal each player seven cards, one at a time.

Within a given time limit, the players try to use any of the "MADS" operations of arithmetic—multiplication, addition, division, or subtraction—to produce three combinations, of two cards each, that equal the same number ($8 \div 4 = 7 - 5 = 1 \times 2$). Or they can use the same or different cards (of the seven they hold) in three pairs whose results make a series ($5 + 4 = 9$; $8 + 2 = 10$; $6 + 5 = 11$; or $3 \times 2 = 6$; $5 + 2 = 7$; $13 - 5 = 8$).

If the player makes the set and first series above, he scores $2 + 2 + 2 = 6$ points for the set and $9 + 10 + 11 = 30$ points for the series, for a total of 36 points. Then he discards one card, replacing it from the top of the deck.

The game continues until all the cards have been drawn or one player has laid down all his cards. Scores for each round are added and 2 points are awarded for each card laid down. High score wins.

Sum Series

Recognizing Arithmetic, Geometric, and Fibonacci Series

Two, three, or four players

Deck of playing cards

Timer

After the cards have been shuffled, ten are dealt, one at a time, to each player.

At a given signal the players try to arrange their cards so that, by means of "MADS" (multiplication, addition, division, and subtraction) they form the terms of an arithmetic series, that is, a series of numbers in which the difference between any two successive terms is the same. For example, 5, 8, 11, . . . , are part of an arithmetic series in which the difference between successive terms is 3.

A player may use the value of a single card as one of the terms in his series. Or to form a term in his series he may multiply or add the numerical values of two or more cards. Or he may divide the value of one card by that of another or subtract a lesser value from a larger one. These arithmetic operations are represented in this game by the players' arranging their cards as in the top illustration on the facing page.

The players work against a time limit. When time is up, the results they have achieved are compared.

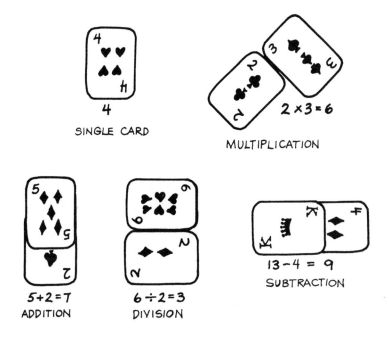

SINGLE CARD

MULTIPLICATION

$2 \times 3 = 6$

$5 + 2 = 7$
ADDITION

$6 \div 2 = 3$
DIVISION

$13 - 4 = 9$
SUBTRACTION

Suppose Player 1 is dealt 6, 2, jack (11), 7, 2, king (13), 3, 1, 5, and 4. He can arrange the cards like this:

The first term in the series is $6 + 2 = 8$. The second term is 11. The third term is $7 \times 2 = 14$. The fourth term is $13 + 3 + 1 = 17$. And the fifth term is $5 \times 4 = 20$. In this series the difference between any two successive terms is 3.

Player 1's score is the sum of the terms in his series plus a bonus of 10 points for having used all ten cards: $8 + 11 + 14 + 17 + 20 + 10 = 80$.

(For the reader's convenience we listed the cards in the order the player used them. He could, of course, use them in a different order.)

If Player 2 is dealt 6, 3, 1, 1, 2, 6, 4, 2, 3, and 10, he can arrange them to make a series with a difference of 2, thus: $6 \div 3 = 2$. $1 + 1 + 2 = 4$. The 6 stands by itself. $4 \times 2 = 8$. The 10 is the last term in the series. The sum of the terms is $2 + 4 + 6 + 8 + 10 = 30$. But from this must be subtracted the value of the unused card, 3. So his score is $30 - 3 = 27$.

Suppose Player 3 receives 5, 2, 10, king (13), 8, 2, queen (12), 7, queen (12), and 10. He can follow $5 + 2 = 7$ with 10, 13, $8 \times 2 = 16$, $12 + 7 = 19$, and $12 + 10 = 22$. His score would be $7 + 10 + 13 + 16 + 19 + 22 = 87$, and the 10-point bonus for having used all the cards makes 97.

High score, after all the cards have been dealt, wins.

A more challenging variation requires the players to form geometric series, that is, a progression of numbers in which the ratio of any two successive terms is the same, such as the series 2, 4, 8, 16, 32, . . . , where each term is the preceding term times 2.

In the most difficult variation, the players produce Fibonacci series, in which each term is the sum of the two preceding terms. For example, in the basic series 0, 1, 1, 2, 3, 5, 8, 13, 21, ... , $1 = 0 + 1$; $2 = 1 + 1$; $3 = 2 + 1$; $5 = 3 + 2$; $8 = 5 + 3$; $13 = 8 + 5$; $21 = 13 + 8$; and so on.

Means and Extremes

Multiplication, Division, Algebraic Equations

Two or more players   Two jokers

Deck of playing cards   Timer

This game is based on the law of proportion. So let us first see how this law operates.

A *ratio* is formed when one number is divided by another — for example, $3 \div 4$ or $(1/2) \div 5$. Another way of writing these and other ratios is to leave out the division line and just insert two vertical dots: 3:4, or (1/2):5. Ratios written this way are read as "Three is to four" and "One-half is to five."

Two equivalent ratios form a *proportion,* for example, 1:2::2:4. This is read as "One is to two as two is to four." In a proportion, the numbers are called *terms.* Thus, in the proportion 1:2::2:4, the 1 is the first term, the first 2 is the second term, the second 2 is the third term, and 4 is the fourth term. The first and the fourth terms of a proportion are the *extremes;* the second and the third terms, the *means.*

To determine whether two ratios are equal and therefore form a proportion, apply the law of proportion: this states that in a proportion the product of the means always equals the product of the extremes. In the proportion 1:2::2:4, the product of the means (2 × 2 = 4) equals the product of the extremes (1 × 4 = 4).

The law of proportion may be applied to many problems in medicine, chemistry, physics, and various trades.

And now we are ready for the game.

Remove all the cards but those with values from 1 through 9. Set aside two jokers, face up. Shuffle the cards and place the deck face down on the table.

The first player takes three cards from the top of the deck and one joker. He makes a row with the joker face up and the other three cards face down, in any order he desires: three cards to the left of the joker; two to its left and one to its right; one to its left and two to its right; or all three to its right. In short, he is free to make the joker the first, the second, the third, or the fourth term in the proportion.

Turning the cards face up, he applies the law of proportion to find the value of the joker.

Suppose his cards are 5, 3, joker, 6. Treating this arrangement as a proportion, he would read it as: 5:3::j:6, that is, "Five is to three as joker is to six." Applying the law of proportion, he multiplies the extremes (5 × 6) and the means (3 × j) and equates them: $3j = 5 \times 6 = 30$. Then $3j = 30$ and $j = 10$. He announces, "The joker is a ten."

The player scores the value of the joker if he solves the equation correctly; if not, he is penalized the value of the correct answer. The first player to score 100 points wins.

The game can be played in another way (with a time limit) that reduces the element of luck and increases the role played by the players' mathematical abilities. Each player can turn up all three cards drawn from the deck, examine them, and decide how to arrange them with the joker to score the maximum number of points.

For example, if a player turns up 5, 2, and 6, he can choose from any of the following three basic patterns:

(1) 5:2::6:j
(2) 2:6::5:j
(3) 6:2::5:j

To be sure, each of these patterns has seven variants, but these do not change the value of j, which is equal to 12/5 in (1), 15 in (2), and 10/6 in (3).

The seven variants of (1) are:

(A) 5:6::2:j
(B) 2:5::j:6
(C) 6:5::j:2
(D) j:2::6:5 $j = 12/5$
(E) j:6::2:5
(F) 6:j::5:2
(G) 2:j::5:6

The seven variants of (2) are:

(A) j:6::5:2
(B) j:5::6:2
(C) 6:2::j:5
(D) 6:j::2:5   } j = 15
(E) 5:j::2:6
(F) 5:2::j:6
(G) 2:6::5:j

The seven variants of (3) are:

(A) 5:6::j:2
(B) 5:j::6:2
(C) 2:6::j:5
(D) 6:5::2:j   } j = 10/6
(E) j:5::2:6
(F) j:2::5:6
(G) 2:j::6:5

So, of course, the player should lay down the second pattern or any of its variants.

If the game is played in this way, a limit should be placed on the time permitted to a player for choosing the best arrangement of the cards.

Root Search

Finding Square Roots

Two players   Deck of playing cards

Paper and pencil

Remove the face cards, leaving the 1 (ace) through 10 in all suits.

The players take turns drawing one card face up until each has five cards. If a card matches in value one previously drawn either by a player or his opponent, it is discarded. The number 10 is counted as 0.

Suppose Player A draws 7, 3, 0, 1, and 8, and Player B draws 2, 4, 5, 6, and 9.

Shuffle the remaining cards with the discards and place them face down. Player A now draws two cards from the top of the deck, say a 7 and a 9, making the number 79. Since the 7 is in A's set of five cards, A scores 1 point. (See score sheet below.) As the 9 is in B's set, he too will score 1 point.

The first player now begins to extract the square root of 79, writing:

$$\overset{\textstyle .}{\sqrt{79.00}}$$

He writes a decimal point after the 9 and also above the line. Since he has used two 0's, and since 0 is in his set of cards, he scores 2 additional points.

Next, he estimates the square root, taking $8 \times 8 = 64$ and $9 \times 9 = 81$. Hence, the square root of 79 must lie between 8 and 9. It is greater than 8 (>8) and less than 9 (<9). So A writes the lower limit, 8, above the line:

$$\overset{\textstyle 8.}{\sqrt{79.00}}$$

As 8 is another number in A's set, A scores 1 point for

the 8. Multiplying $8 \times 8 = 64$, he writes 64 under the 79:

$$
\begin{array}{r}
8. \\
\sqrt{79.00} \\
64
\end{array}
$$

Since the 6 and the 4 in 64 are B's cards, B gets 2 points.

Now A, as the next step in the calculation, subtracts 64 from 79, giving 15. This is written, and the two zeroes are brought down as shown:

$$
\begin{array}{r}
8. \\
\sqrt{79.00} \\
\underline{64} \\
15\ 00
\end{array}
$$

A scores 1 point for the 1 in 15, and B scores 1 point for the 5. The next trial divisor is formed by doubling the part of the root found in the previous step: $8 \times 2 = 16$. The result should now be written down with a dash to its right as illustrated. The 1 and the 6 in 16 are the first two digits of a three-digit divisor of 1500, the third digit of which is now to be found. The dash is written for the unknown digit:

$$
\begin{array}{r}
8. \\
\sqrt{79.00} \\
64 \\
16\_\overline{)15\ 00}
\end{array}
$$

A scores 1 point for the 1 in 16, and B scores 1 point for the 6.

To determine the last digit of the trial divisor, which will be matched above the line, A may try dividing 1500 by 160. $1500 \div 160 = 9$. But $169 \times 9 = 1521$, which is greater than 1500 and won't do. He tries 8: $168 \times 8 = 1344$, which is less than 1500. So he writes in the 8 after the decimal place above the line and also as the last digit of the divisor, inserts the product of 168 and 8 (1344) under 1500, and subtracts: $1500 - 1344 = 156$.

The completed work, to one decimal point, which is as far as we shall carry the computation here, then looks like this:

$$
\begin{array}{r}
8.8 \\
\sqrt{79.00} \\
64 \\
\overline{168\,)\,15\ 00} \\
13\ 44 \\
\hline
1\ 56
\end{array}
$$

The 8 will count as 1 point for Player A, and he will score 2 points more for the 1 and the 3 in 1344. Player B will score 2 points for the two 4's in 1344. The 1 in 156 will count 1 point for Player A, and the 5 and the 6 will count 2 points for Player B. Thus, the score sheet will look like this:

	Player A	Player B
7, 9	1	1
0, 0	2	
8	1	
6, 4		2
1, 5	1	1
1, 6	1	1
8, 8	2	
1, 3, 4, 4	2	2
1, 5, 6	1	2
	11	9
Square Root:	8.8	
Total:	19.8	9

Player A has added the value of the square root to his score, since he did the computations. He leads, 19.8 to 9.

Player B now draws two cards and follows the same procedure with a new square root. The player with the higher score after each has extracted three square roots wins.

The game can be made more difficult by adding decimal places to the square root. Even more challenging is extracting square roots of three-digit numbers; in this case, the number must first be separated into 2-digit "periods," marking off from the decimal point. After the completion of this step, the same procedure as previously outlined may be followed, treating the first period as the number whose square root must initially be approximated.

I Doubt It

Addition, Subtraction, Multiplication, Division, Squares, Square Roots

Two, three, or four players   Timer

Deck of playing cards

As in previous games, the jack is worth 11; the queen, 12; and the king, 13.

Shuffle the cards and deal out four face down on the table. This is the reserve set.

Now deal the remaining cards, one at a time, to the players.

Using one, two, three, or four cards, the first player must, within a given time, perform on their numerical values any of the basic arithmetical operations—addition, subtraction, multiplication, or division, singly or in combination—to produce a sum, a remainder, a product, or a quotient of 1. He does this to one side, concealing his work from the eyes of his opponents. When he has worked out his solution, he places the cards face up before him and says, "One."

Any opponent may challenge him by saying, "I doubt it." Then the first player must state what operations were used to produce the number 1. If he is correct, the challenger has to pick up the cards on the table (except the reserve set). If not, the player challenged must pick them up.

For example, if the first player used king, queen, jack, and 10, he could subtract the queen from the king $(13 - 12)$, subtract the 10 from the jack $(11 - 10)$, and multiply the results: $1 \times 1 = 1$.

The second player, using the same means, must produce the number 2 within the allotted time. Using 8, 4, 6, and 5, for instance, he could calculate as follows: $(8 \div 4)(6 - 5) = 2 \times 1 = 2$. However, all he would do would be to put down the four cards face up and say, "Two." He could then meet any challenge.

The third player tries to reach 3 in the same way. As the count increases, squares and square roots can also be introduced into the calculations.

If a player cannot reach his required value in the allotted

time with the cards he has in his hand, he may draw one card from the reserve set of four on the table. If this does not help, he may try a bluff, saying whatever number he is supposed to reach and laying his cards face up on the table. Of course, if he is challenged, he will have to pick up the cards again, along with all the unchallenged cards that have been laid down.

A player can plan his strategy by calculating in advance the value he has to obtain when his turn comes next. For example, if he has to produce a 3 now, and there are four players in the game, he can tell that he will have to produce a 7 in the next round. In this way, he can prepare to make the best use of his hand. This knowledge may help him to use as many cards as possible (up to four) in each round.

The first player to use up all his cards wins.

To add excitement, the time limit can be shortened for each succeeding round.

Prize Contest

Multiplication, Calculation of Percentages

Two or more players   Graph paper

Paper and pencil   Deck of playing cards

To spur the collection of real estate taxes, some cities offer prizes to their employees if weekly goals are met.

The players in this game imagine that they are competing for these prizes.

Shuffle the cards and place them face down on the table. The jack is valued at 11; the queen, at 12; and the king, at 13.

The first player draws two cards. The higher card represents thousands of dollars of assessed valuation; the lower, hundreds of dollars. Thus, if he draws a queen (12) and a 3, the assessed valuation of the property to be taxed is $12 \times 1,000 + \$3 \times 100 = \$12,300$.

The property tax rate is fixed for each game in advance. For example, a tax rate of $5.96 per $100 (or $59.60 per $1,000) of assessed valuation is equivalent to a rate of 5.96 percent.

With this rate, suppose that the first player draws a jack and a 5. The assessed valuation of the property to be taxed is $11,500. To calculate the tax, he multiplies the assessed valuation by 0.0596. The answer is $685.40. Rounded off to the nearest $100, this is $700, which is entered to his credit on the thermometer graph, as shown.

If the second player draws a queen and a jack, the assessed valuation is $12,000 + \$1,100 = \$13,100$. The tax would be 5.96% of this amount, or $\frac{5.96}{100} \times \$13,100 = \$131 \times 5.96 = \$780.76$. Rounded off to the nearest $100, this comes to $800, which is credited to the second player and entered on the graph.

The two players together have so far collected $1,500 to-

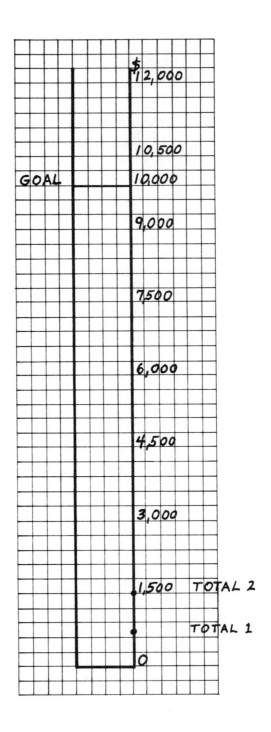

GOAL

$12,000

10,500

10,000

9,000

7,500

6,000

4,500

3,000

1,500   TOTAL 2

          TOTAL 1

0

66

ward the city's goal. If the goal is $10,000, the game continues until the players reach or exceed it. The player who goes to or over the top gets a prize of $1,000, which is added to his score. Also, he can double the assessed valuation on which he calculates the tax if the two cards he draws are of the same suit. For example, if he draws the 8 and the 10 of clubs, the total valuation is $10,800 × 2 = $21,600. Multiplied by 0.0596, this results in a tax of $1,287.36, or $1,300 rounded off to the nearest $100.

The player with the highest total score (taxes plus $1,000 prize if won) wins the game.

For each new game the tax rate should be changed.

If a player makes an error in the calculation of the tax, any other player can challenge him. If the challenger then correctly calculates the tax, he is credited with it.

Various Areas

Calculating Areas of Rectangles, Triangles, Parallelograms, and Circles

Two players   Deck of playing cards

Paper and pencil

In this game the jack represents a triangle; the queen, a rectangle; and the king, a parallelogram.

Divide the cards into two packs, one consisting of the picture cards (jack, queen, and king), and the other of the rest of the deck. Lay the packs face down, side by side, on the table.

The first player draws one card from the pack of picture cards and two from the other pack. He turns the three cards over, sketches the geometric figure represented by the picture card, and calculates its area, using the numbers on the other two cards.

The area of a rectangle is $A = bh,$ where b is the length of one side and h is the height of a side that intersects it. To calculate the area of a triangle, use the formula: $A = \dfrac{bh}{2},$ where b is the length of one side and h is the height of a line drawn at right angles to that side from the point where the other two sides intersect, as in the diagram. The area of a parallelogram is $A = bh,$ where b is the length of a side, and h is the height of a line drawn at right angles to that side and extending to the opposite side, as illustrated.

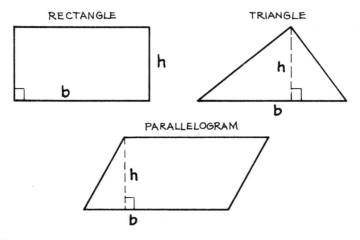

RECTANGLE   TRIANGLE

h   b   h   b

PARALLELOGRAM

h   b

If Player 1 draws a queen, an 8, and a 6, he sketches a rectangle with sides of 8 and 6 inches (or feet). Using the formula, he multiplies: $8 \times 6 = 48$ square inches or square feet. His score is 48 points.

If Player 2 draws a jack, a 5, and a 7, he sketches a triangle, and applies the formula thus: $\dfrac{7 \times 5}{2} = \dfrac{35}{2} = 17\ 1/2 = 18$ (rounded up). He scores 18 points.

The game continues until all the cards from the picture pile are gone. High total score wins.

To make the game harder, the jack or some other picture card can represent a circle. If a player draws a jack, he adds the two numbered cards to determine the radius of the circle, that is, the distance from its center to any point on its circumference. The diameter of a circle is twice the length of its radius, and the ratio of the length of the circumference to the diameter is a constant denoted by the Greek letter π (pronounced "pi"). This ratio, which is the same for all circles, is *roughly* equivalent to 22/7. The formula for calculating the area of a circle is: $A = \pi r^2$. So if a player draws a 3 and a 7 with a picture card representing a circle, the area of the circle is: $A = \pi(3 + 7)^2 = \dfrac{22(10)^2}{7} = \dfrac{2200}{7} = 314.3$ or 314 (rounded to the nearest unit).

Play Room

Calculating the Area of an Equilateral Triangle

Two or more players   Deck of playing cards

Paper and pencil

The room for play here is the area enclosed by an equilateral triangle. To calculate this area, apply the formula:

$A = \dfrac{s^2}{4}\sqrt{3}$, where A is the area and s is one of the three equal sides.

Remove the picture cards from the deck, shuffle the deck thoroughly, and lay the cards face down on the table.

The first player draws a card from the top of the deck. Counting 1 for clubs, 2 for diamonds, 3 for hearts, and 4 for spades, he calculates the area of his triangle by multiplying the value of the suit he has drawn by the number shown on the card and treating the product as the length of a side.

Thus, if he draws the 3 of spades, he multiplies 3 by 4 (the value of spades): $s = 3 \times 4 = 12$. He then applies the formula: $A = \dfrac{12^2\sqrt{3}}{4} = \dfrac{144\sqrt{3}}{4} = 36\sqrt{3}$. (The answer, his score, may be left in radical form.)

Each player follows the same procedure, scoring the value of the area of his triangle. The first player to attain a total score of $200\sqrt{3}$ or more is the winner.

3

Games with Dice
and Dominoes

MANY OF THE GAMES that people of all ages like to play involve rolling dice or matching dominoes. The pips on either denote numbers from 1 through 6 that can be used in many different combinations, as in the games in this chapter, to provide practice in basic arithmetical skills and in computations requiring algebraic formulas, radicals, exponents, and some elementary geometric theorems. If pencil and paper or other materials are also required for any game, this information will be found under its title.

71

Fraction Race

Addition of Fractions

Two or more players   Red and white dice

Paper and pencil   Colored disks

Graph paper

In this game the number on the red die represents the numerator of a fraction, and the number on the white die, the denominator. Thus a combination of white 2 and red 3 is equivalent to 2/3, while a combination of white 3 and red 2 is read as 3/2.

Since the figures on each die run from 1 through 6, the least common denominator is 60. Accordingly, each square on the graph paper used to chart the race counts as 1/60. So 1/6 is equivalent to ten squares; 1/5, to twelve squares; 1/4, to fifteen squares; 1/3, to twenty squares; 1/2, to thirty squares; 1 (as in 3/3 or 5/5), to sixty squares; 1 1/2, (e.g., 3/2), to ninety squares; etc. The chart should be marked accordingly, as shown (in part) in the illustration.

Now the players take turns rolling the dice. Whatever fraction is rolled is recorded on the graph with a differently colored marker for each player.

Suppose that Player A rolls white 1, red 2, which is equivalent to 1/2. If he has a red disk, he places it on the point marked 1/2 on the chart, thirty squares from 0. If Player B rolls 2/6 = 1/3, he places his disk — let us say it is

72

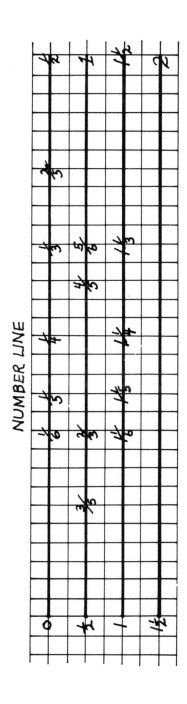

NUMBER LINE

73

blue—on the twentieth square from 0. Player C, with a yellow disk, might roll $4/2 = 2$ and would then place his marker on the 120th square.

Now if Player A, on his second turn, rolls white 2, red 1, or $2/1 = 2$, he will move $2 \times 60 = 120$ squares beyond his original position at 1/2. $30 + 120 = 150$. He can say: $2 + 1/2 = 2\ 1/2$ and place his red disk on the 150th square, which is equivalent to 2 1/2.

If Player B now rolls white 5, red 4, or $5/4 = 1\ 1/4$, he can say: $1\ 1/4 + 1/3 = 1\ 15/60 + 20/60 = 1\ 35/60$ and can place his blue disk on the 95th square. (It is, of course, possible for more than one marker to occupy the same square.)

The first player to advance his marker to the finish line, equivalent to 600 squares or 10 units, is the winner.

Decimal Roll

Decimal Operations

Two players   Timer

Red and white dice

The numbers on the red die in this game represent the numerators of fractions, and those on the white die, the denominators.

Player A rolls the dice to get a fraction. For example, if he

74

rolls red 3, white 4, his fraction is 3/4. Within a given time limit, he must convert this fraction into its decimal equivalent, 0.75. If he rolls red 6, white 5, his fraction is 6/5 = 1.2.

A player scores 1 point for correctly stating, within the time limit, the decimal value of the fraction he rolls. He adds this point to the decimal to get his score for the roll. Thus, if he correctly converts 5/4 to 1.25, his score is 1 + 1.25 = 2.25; if he correctly converts 5/6 to 0.833, his score is 1.833.

The players take turns rolling the dice. The first player whose accumulated score, rounded off to one decimal place, is 25 or more wins.

Since there are a limited number of possible fractional combinations in this game, the same fractions will recur. Soon the players should be able to recall their decimal equivalents quickly and become familiar with them.

The game can be made more difficult by reducing the time limit.

MADS

Multiplication, Addition, Division, Subtraction

Two or more players   Two dice

Each player takes his turn rolling a pair of dice once. He

then uses "MADS" on the result; that is, he performs the operations of multiplication, addition, division, and subtraction, in that order, to arrive at his score.

Suppose the first player rolls 4 and 6. He arranges these figures with the larger one first: 6, 4. Then he:

Multiplies $6 \times 4 =$	24
Adds $6 + 4 =$	10
Divides 6 by 4 to get $6/4 = 1\ 1/2$, rounded to	2
Subtracts $6 - 4 =$	2
Total score:	38

If the second player rolls 5 and 3, he scores $5 \times 3 = 15$, $5 + 3 = 8$, $5 \div 3 = 1\ 2/3$ rounded to 2, and $5 - 3 = 2$, for a total score of 27 points.

The first player to reach a total of 200 points wins.

Sum Fractions

Fraction Operations

Two or more players   Three dice

Paper and pencil   Timer

When three dice are rolled, we can call the numbers that turn up a, b, and c. They can form six fractions: a/b, a/c, b/a, b/c, c/a, and c/b.

Each player rolls the three dice and scores the sum of the six fractions that can be formed from the numbers turned up, provided he can write out the addition of the fractions correctly within the given time limit.

For example, if Player 1 rolls 2, 5, and 6, the six fractions these numbers form are 2/5, 2/6, 5/2, 5/6, 6/2, and 6/5. There are various ways of adding these fractions. The player can convert them to fractions with their least common denominator, which is 30. The numerators then become 12, 10, 75, 25, 90, and 36. The sum of the numerators is 248. $248 \div 30 = 8\ 8/30$, which rounds off to 8. Or the player can add the fifths, sixths, and halves separately: $2/5 + 6/5 = 8/5$; $2/6 + 5/6 = 7/6$; $5/2 + 6/2 = 11/2$. Then $8/5 + 7/6 + 11/2 = 1\ 3/5 + 1\ 1/6 + 5\ 1/2 = 7 + 18/30 + 5/30 + 15/30 = 8\ 8/30$, or 8.

Now, if Player 2 rolls 1, 2, and 3, his six fractions would be: $1/2 + 1/3 + 2/1 + 2/3 + 3/1 + 3/2$. The least common denominator here is 6. The numerators become $3 + 2 + 12 + 4 + 18 + 9$. Their sum is 48; and $48/6 = 8$, which is also the sum of 4/2, 3/3, and 5/1. (A third method of adding is to convert to decimals and add them: $0.5 + 0.33 + 2.0 + 0.67 + 3.0 + 1.5 = 8$.)

Scoring is cumulative.

If a player makes an error in addition or cannot calculate his score within the time limit, he scores 0 and loses his turn.

The winner is the player whose cumulative score first reaches 50 points or more.

Visible Divisibles

Addition, Division

Two or more players   Timer

Dominoes

This game teaches the players a mathematical shortcut: instant divisibility by 11.

A three-digit number is divisible by 11 if the middle digit is the sum of the other two digits. For example, the middle digit, 7, of the number 374 is the sum of the other two digits, 3 and 4. If you now divide 374 by 11, the answer is 34, a number consisting of the other two digits in the order in which they appear in the dividend.

Mix up the dominoes thoroughly and place them face down on the table. The first player turns up two dominoes — say, with the combinations 5, 3 and 6, 2. Within a given time limit, using the two dominoes, he is to form a number divisible by 11, if possible. In this case, $5 + 3 = 8$. Placing the 8 between the 6 and the 2, we get 682, in which the middle digit is the sum of the other two. 682 is divisible by 11; the quotient is 62. If the player works this out correctly in time, he earns the sum of the three digits: $6 + 8 + 2 = 16$. (If he placed the 8 between 2 and 6, 5 and 3, or 3 and 5, he would also score 16.)

Another two dominoes are turned up to replace the ones used. The second player turns up 4, 3 and 5, 1; he can do nothing and scores 0. If he had turned up 2, 5 as his second

78

domino, he could have added these to make 7 and placed the 7 between the 4 and the 3 to make 473. Divided by 11, this is 43. He would have scored $4 + 7 + 3 = 14$.

The game continues until all the dominoes are used up. The player with the high score wins.

On Your Mark!

Addition and Multiplication of Decimals

Two or more players

Red and white dice

Paper and pencil

Graph paper

In this game the numbers on the red die are to be considered whole numbers (integers), and those on the white die are to be treated as one-place decimals. A combination of red 3 and white 4 would be read as 3.4, and a combination of red 1 and white 6 would be equivalent to 1.6.

Each player, in turn, rolls the dice twice and adds the results of the two throws to determine his score. Thus, red 3, white 4, and red 2, white 5, score $3.4 + 2.5 = 5.9$. The result, rounded to the nearest whole number (in this case

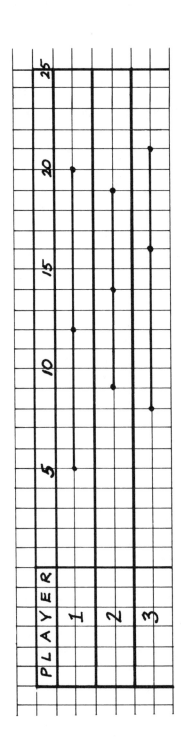

80

6) or to the next higher integer if the number ends in .5, is recorded on a scorecard like the one shown here.

Each player has played three rounds, cumulating his scores by connecting them with a line on the scorecard. The player who first reaches the finish line — in this game set at 25 — wins.

After a few rounds of adding the die number equivalents, the players can try multiplying them. In that case, the finish line should be set higher, or the winner can be the player who has the highest total score after, say, five turns for all.

Weigh-in

Addition, Subtraction, Multiplication, Division

Two or more players   Two dice

With five weights — 1 pound, 2 pounds, 4 pounds, 8 pounds, and 16 pounds — a storekeeper can weigh anything that weighs exactly 1 pound through 31 pounds by placing the object to be weighed on one pan of a balance and from one through five weights on the other.

The players in this game imagine that they have these five weights at their disposal. The object is to see how much each player can weigh with each throw of the dice when he performs "MADS" (multiplication, addition, division, and subtraction) on the two numbers thrown.

Suppose the first player throws two 6's. Multiplying them, he gets 36. This cannot be weighed with the five weights at his disposal. Adding the two 6's, he gets 12 — the sum of the 8-pound and 4-pound weights. Dividing one 6 by the other, he gets 1, which matches the 1-pound weight. Subtracting one 6 from the other, he gets 0, which cannot be weighed. So all that he can weigh is $12 + 1 = 13$. But for rolling a double number, he scores a 5-point bonus. Thus, his score is $13 + 5 = 18$. Of course, if he makes an error, he scores nothing.

Each player follows the same procedure. The first player to attain a cumulative score of 150 pounds or more is the winner.

Eke-Weight

Addition, Subtraction, Multiplication, Division, Equations

Two or more players   Timer

Two dice

This is a variation of Weigh-in, but more challenging.

Each player imagines that he has just four weights — 1, 3, 9,

82

and 27 pounds. With these he can weigh anything from exactly 1 pound through 40 pounds.

Balancing 4 pounds is like the procedure in Weigh-in: the 1- and 3-pound weights in one pan balance the object. But how do you balance a 2-pound object?

That's where the complication comes in. The weights can now go on the other pan as before, or some weights can go *on the same pan as the object*. Thus the 1-pound weight is put on the pan with the 2-pound object and balanced by the 3-pound weight on the other pan.

The name of the game is derived from the player's having to *eke* out an *equality*. Within a given time limit he has to combine one or more of these weights with the four results of using "MADS" on the two numbers that turn up when he throws the dice to "eke-weight" the figures arrived at, that is, to get equal weights on both sides of a scale, so that they balance perfectly. In short, he has to set up a series of equations.

Suppose Player A rolls a 5 and a 3. Using "MADS," he first multiplies: $5 \times 3 = 15$. How does he demonstrate that he can "eke-weight" this with one or more of the four weights? By putting the 15 pounds with the 3-pound and the 9-pound weights on one side of the scale, and placing the 27-pound weight on the other side. The equation will then be: $15 + 3 + 9 = 27$.

He next adds the 5 pounds and the 3 pounds: $5 + 3 = 8$. He can use this with the 1-pound weight and against the 9-pound weight to produce the equation: $8 + 1 = 9$.

He then divides the 5 pounds by the 3 pounds: $5 \div 3 = 1$ 2/3 or 2 (rounded off). Using 2 pounds with the 1-pound and against the 3-pound weight, he gets the equation: $2 + 1 = 3$.

Finally, he subtracts the 3 pounds from the 5 pounds: $5 - 3 = 2$. The equation will be the same as the one used when he did the division.

Player A can count the 2 pounds only once, since he calculated the equation just once. Therefore, his score is $15 + 8 + 2 = 25$, provided he does all the calculations correctly within the allotted time.

The first player with a cumulative score of 150 points or more is the winner.

Ups and Downs

Addition, Subtraction, Multiplication, Division, Positive and Negative Numbers, Plotting Points

Two or more players   Two dice

Pencil   Timer

Graph paper

In this game, no matter what direction you move in, up or down, you are bound to approach your goal. But you must move fast! Within the time limit, that is.

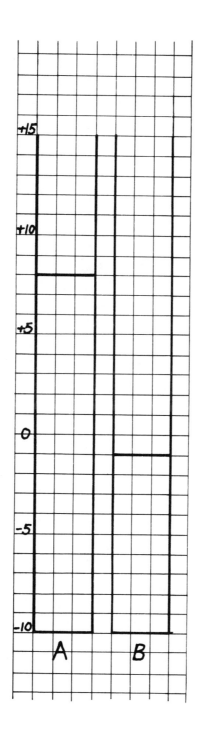

Player A throws the dice. First, he performs "MADS" (multiplication, addition, division, and subtraction) on the two numbers that turn up. Thus, if he throws 3 and 5, he gets $3 \times 5 = 15$; $3 + 5 = 8$; $5 \div 3 = 1$ 2/3, or 2 (rounded off); and $5 - 3 = 2$.

Now, he adds the four results: $15 + 8 + 2 + 2 = 27$. Placing the smaller of the two numbers he has thrown (3) in front of the larger (5), he gets 35. From this he subtracts 27. $35 - 27 = 8$. This is his score, which he is allowed to enter on the ladder graph as shown, if he has done all his calculations correctly within the time limit.

Suppose Player B rolls 2 and 6. $2 \times 6 = 12$; $2 + 6 = 8$; $6 \div 2 = 3$; $6 - 2 = 4$. Then $12 + 8 + 3 + 4 = 27$. The 2 placed before the 6 makes 26. $26 - 27 = -1$. This negative number, if arrived at correctly within the allotted time, is likewise entered on the ladder graph, as illustrated. Each player's ladder is designated by a letter.

A player wins by reaching the top of the ladder ($+15$) or by hitting the bottom (-10).

Zero In!

Addition, Subtraction, Multiplication, Division, Positive and Negative Numbers

Two or more players   Red and white dice

Paper and pencil   Timer

Each player writes down the following pattern: (__) × (__) + (__) ÷ (__) − (__).

Using the white die for positive numbers and the red die for negative numbers, the first player rolls the dice five times, adds separately the positive and negative numbers turned up by each throw, and enters the sum, whether positive or negative, in the spaces of the pattern on his paper, proceeding from left to right.

Thus if he rolls white 5, red 2; white 4, red 3; white 5, red 1; white 2, red 3; and white 4, red 6, the respective sums are $5 - 2 = 3$; $4 - 3 = 1$; $5 - 1 = 4$; $2 - 3 = -1$; $4 - 6 = -2$. These are entered in the spaces of the pattern as follows: $(3) \times (1) + (4) \div (-1) - (-2)$. It is then necessary to follow the formula for order of operations: in the case of a series of operations not divided into mathematical phrases by brackets, MADS must always be performed in the order MDAS, otherwise known as My Dear Aunt Sally. That is, multiplication and division must be finished before addition and subtraction can be begun. Thus, in our example, $3 \times 1 = 3$; $4 \div (-1) = -4$; $3 + (-4) = -1$; $-1 - (-2) = 1$. This is the first player's score, if he has calculated it within the allotted time (which began at the moment he picked up the dice).

The score of each player is kept cumulatively. If a player's first four turns score −6, −8, +3, and +4, his cumulative scores are −6, −14, −11, and −7.

The object of the game is to Zero In by being the first to attain a cumulative score of 0.

If ten rounds pass and no player has reached 0, the player whose cumulative score is closer or closest to zero wins. A player with +3 would lose to a player with −2.

Palindromes

Addition

Two or more players   Timer

Three dice

This game requires really rapid addition.

A palindrome is any number that reads the same from right to left as from left to right, like 646, 8228, and 47,274.

Every number can be used to produce a palindrome. Just reverse the order of its digits and add the reversed number to the original number. If the sum is not a palindrome, reverse the sum and add again, and continue in this way until you form a palindrome.

For example, if you begin with 127, the reversed number is 721. $127 + 721 = 848$, a palindrome reached in just one step.

If you begin with 517, add 715. The result, 1232, is not a palindrome. Reverse the digits in 1232 and add: $1232 + 2321 = 3553$, a palindrome reached in two steps.

Or try starting with 678. Then $678 + 876 = 1554$; $1554 + 4551 = 6105$; $6105 + 5016 = 11,121$; and $12,111 + 11,121 = 23,232$, a palindrome reached in four steps.

Each player throws three dice and then tries to make a palindrome from the numbers on the dice in as few steps as possible. If two players take the same number of steps, the faster one wins. Or without a timer you can score the game like this: 5 points plus the sum of the digits for a palindrome produced in one step; 3 points plus the sum of the digits for a palindrome formed in two steps; and just the sum of the digits for a palindrome arrived at in more than two steps.

If a player throws three numbers that are the same, like 666, he has already produced a palindrome and need go no further. In that case, he scores 15 points plus the sum of the digits — that is, $15 + 18 = 33$.

If a double number turns up, the player can arrange the order of the digits to form a palindrome immediately. Thus, two 2's and a 1 form 212. For producing this a player scores 10 points plus the sum of the digits — in this case, $10 + 5 = 15$. If he failed to see this possibility and began with the number 221 or 122, he could form a palindrome in one step: $221 + 122 = 343$. He would then score 5 points plus the sum of the digits, that is, $5 + 10 = 15$, just as if he had formed the palindrome immediately. So a penalty of 3 points is imposed for his failure to produce the palindrome at once.

Obviously the order in which a player arranges the three numbers on the dice can affect his score because it will determine how many steps he will need to produce a palindrome. For example, if he throws 3, 2, and 6, he can form 236, 263, 326, 362, 623, or 632. However, these six possibilities can be reduced to just three: 236, 263, and 326, since the other three numbers, being simple reversals of the digits of the first three, produce the same scores. $236 + 632 = 868$. Here the palindrome is achieved in just one step. The player can thus score $5 + 22 = 27$. $326 + 623 =$ 949. Here too, only one step is needed, and the score is the same: 27. But 263 is an inferior choice: $263 + 362 = 625$, which is not a palindrome. Proceeding, we get $625 + 526 = 1151$ and finally $1151 + 1511 = 2662$, a palindrome reached in three steps. The score is only $2 + 6 + 6 + 2 = 16$.

Accordingly, the player will have to do some fast figuring to choose the best arrangement. The time limit can be made as short as necessary to prevent a leisurely comparison of all three possibilities.

The first player to attain a total score of 100 points or more wins.

Building Blocks

Multiplication

Two players   Dominoes

Mix the dominoes thoroughly and place them face down on the table.

Player A turns up two dominoes. Let us say he draws 6, 3, and 5, 6. Since he has two matching numbers (the two 6's), he begins to build a column with them by placing one above the other as shown. With a column of two matching numbers, he scores double their value: $6 \times 2 = 12$.

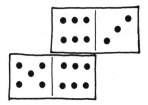

His opponent now turns up two dominoes. If he draws 3, 2, and 1, 4, he cannot build with them a column of his own, and he is never allowed to build on Player A's column. He sets the two dominoes aside for possible later use, and scores nothing.

After the first draw the players turn up only one domino at a time.

Suppose Player A now turns up 1, 6. He continues to build up his column of 6's:

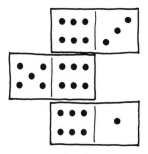

The column of three 6's scores $6 \times 3 = 18$. He cumulates his score, which is now $12 + 18 = 30$.

If Player B next turns up 2, 4, he can build two columns with this and the two dominoes that he set aside:

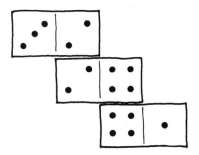

He scores $2 \times 2 = 4$ and $4 \times 2 = 8$. $4 + 8 = 12$.

The game proceeds until all the dominoes are used up. The player with the higher score is the winner.

Leftovers

Fraction Operations

Two or more players   Timer

Dominoes

After mixing the dominoes, place them face down on the table. Then expose three tiles. Let us say that these are 3, 5; 4, 2; and 1, 6.

The first player now draws one tile. Suppose it is 6, 2. He

pairs it with one of the exposed tiles as shown and forms a fraction with the "leftovers," that is, the two numbers that do not match, placing the smaller over the larger. His score

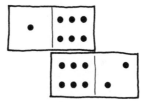

is 1/2. But if he had not been in such a hurry, he might have scored higher, like this:

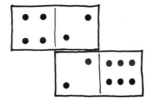

Here the leftovers are 4 and 6. 4/6 = 2/3, which is greater than 1/2. His decision must be made within a given time limit.

The next player may turn up 6, 5. He can pair this with 3, 5. The leftovers are 3 and 6, or 3/6. His score is 1/2.

The game continues in this way, with each player trying to make the most advantageous move within the time limit.

If a player forms a block of three dominoes, the fraction is formed by making the larger number the numerator and the smaller the denominator. In the combination illustrated, the leftovers are 4 and 3. 4/3 = 1 1/3.

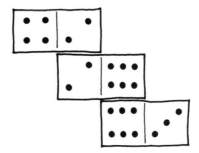

If a block of four dominoes is formed, the two possible fractions formed by the leftovers are added. Accordingly, in the arrangement shown, the leftovers 4 and 1 can be made into 1/4 and 4/1. 1/4 + 4/1 = 4 1/4.

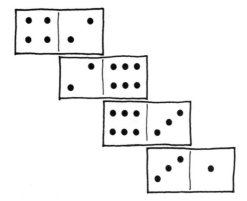

If a block of five dominoes is formed, the larger of the two possible fractions should be multiplied by 2.

Scoring is cumulative. After all the dominoes are used up, high score wins.

94

Beehive Squares

Squaring Numbers, Addition

Two or more players   Two dice

Paper and pencil   Colored disks

Each player prepares his own beehive, consisting of a grid of hexagons, as illustrated.

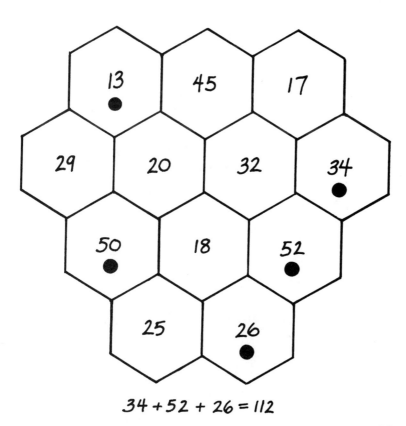

$$34 + 52 + 26 = 112$$

He then enters twelve different numbers in the honeycombs, selecting them from this list: 2, 5, 8, 10, 13, 17, 18, 20, 25, 26, 29, 32, 34, 37, 40, 41, 45, 50, 52, 61, and 72.

These numbers are the possible sums of the squares of the numbers turned up when dice are thrown. Thus, if two 2's are thrown, and if each 2 is squared, the result is $4 + 4 = 8$. If a 6 and a 5 are thrown, the sum of their squares is $36 + 25 = 61$.

Each player in turn rolls the dice and adds the squares of the numbers that come up. If he finds the sum in any of his honeycombs, he places one of his distinctively colored disks on it. The other players do the same in their honeycombs.

The object of the game is to get disks on three contiguous honeycombs. As soon as a player does this, he yells out, "I'm stung!" He scores the sum of the three contiguous boxes. Then a new round is begun with different numbers in each player's beehive. The player who first attains a total score of 200 or more wins.

Metric Crossout

Addition, Multiplication, Division, Metric Conversions

Two players   Two dice

Paper and pencil   Timer

Each player writes down the numbers 2, 3, 4, 5, 6, 7, 8, 9, 10, 11, and 12. These numbers represent 2 pounds, 3 pounds, etc.

Each player, after rolling the dice, must, within a given time, convert the sum of the numbers he rolls from pounds into kilograms, accurate to one decimal place. (A Table of Metric Equivalents is provided in the Appendix.) Thus, if a player rolls 4 and 3, he must convert 7 pounds to kilograms. The answer is 3.2 kg. If he performs this calculation correctly and in time, he crosses out the 7 on his paper and gets another chance to roll the dice. He continues until he either gives an incorrect answer or rolls a combination with a sum already crossed out on his paper. The next player then throws the dice.

The game continues until one player has crossed out a total of 40 pounds (or more). He is the winner.

After a few rounds the conversion may be reversed. The numbers from 2 through 12 could represent kilograms to be converted into pounds. (See Appendix again.) The problem then becomes one in multiplication.

In another variation of this game, the numbers represent meters and are to be changed to inches (to the nearest inch). The problem is again one of multiplication. In this variation each player rolls once. If he can change his correct answer in inches to feet, he is entitled to an additional roll. For example, if he rolls 4 and 1, he first must convert 5 meters to inches. A meter is approximately 39.37 inches; so $5 \times 39.37 = 196.85$. If the player gives the rounded answer, 197 inches, within the time limit, he crosses out 5 on his chart. Now he can try to convert this figure into feet, to one decimal place. Since there are 12 inches in a foot, he must divide 197 by 12. The answer is 16.4 feet. If he performs this calculation correctly, he is entitled to an additional roll. If he wants a second additional roll, he must convert 16.4 feet into yards. As there are 3 feet in a yard, he must divide 16.4 by 3. The answer is approximately 5.5 yards. If he cannot perform the calculations in time and correctly, he passes his turn.

The same variation is playable in converting liters to quarts and gallons, and vice versa.

Area Roll

Calculating the Area of a Rectangle and of a Square

Two players   Two dice

Paper and pencil

The area of a rectangle is the product of the length of two adjacent sides. The area of a square is the square of the length of one side.

The object of this game is to calculate these areas and add them to attain a total higher than one's opponent's.

First write down the following figures: 1^2, 2^2, 3^2, 4^2, 5^2, 6^2, 1×2, 1×3, 1×4, 1×5, 1×6, 2×3, 2×4, 2×5, 2×6, 3×4, 3×5, 3×6, 4×5, 4×6, 5×6.

The first player rolls the dice and multiplies the two numbers that turn up — say, $4 \times 3 = 12$. This is the area of a rectangle 4×3.

He then crosses out of the list of figures any that equal 12 — for example, 4×3 or 6×2 or $3^2 + (1 \times 3)$ or $2^2 + (2 \times 4)$, etc.

Now the other player rolls the dice and multiplies, say, $5 \times 5 = 25$. This is the area of a square 5×5. He then crosses out any of the remaining figures that can be made to equal 25 — for example, 5^2 or $(4 \times 5) + (1 \times 5)$, etc.

The game continues until all the numbers are crossed out. If a player cannot cross out any figures after throwing the dice, he passes, and his opponent rolls the dice. Each player scores the value of the area he has crossed out. If he is unable to cross out any area, he scores 0. If a player rolls two equal numbers — say, 5 and 5, or 3 and 3 — to make a square, he gets a bonus roll of the dice.

The player with the higher total score when all the numbers have been crossed out is the winner.

Area Varier

Calculating the Area of a Circle

Two or more players   Two dice

Paper and pencil

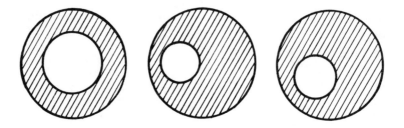

Each of the figures illustrated consists of two circles, one within the other. When the smaller circle has the same center as the larger, the circles are said to be *concentric*, as they are in a ring. Otherwise, they are *eccentric*.

The object of the game is to calculate the area that lies between the circumference of the large circle and that of the small circle. In the illustration, this area is shaded in all three examples.

The area required is simply the area of the large circle minus the area of the small circle. Mathematically, we can express this as: $A = \pi R^2 - \pi r^2$, where R is the radius of the large circle and r is the radius of the small circle.

Player 1 rolls the dice. The larger number rolled is R, and the smaller number is r. So if he rolls 5 and 3, 5 inches is the radius of the large circle, and 3 inches is the radius of the small circle.

Substituting these values in the formula, we get $A = \pi(5^2) - \pi(3^2) = 25\pi - 9\pi = 16\pi$. We leave the answer expressed in terms of π. Player 1's score is 16π.

If a player throws two equal numbers — for example, two 5's or two 3's — his score is 0.

100

The winner is the player who first attains a total score of 100π.

Tangent Dash

Applying the Pythagorean Theorem, Finding Squares and Square Roots

Two or more players   Two dice

Paper and pencil

A *tangent* to a circle is a line that can touch only one point on its circumference. In the figure, the tangent is marked T. If we now draw a straight line S from the center of the circle to intersect with the tangent T, and if we draw the radius R from the center of the circle to the point where the tangent touches the circumference, we form a right triangle, because the angle formed by T and R is always 90° or a right angle.

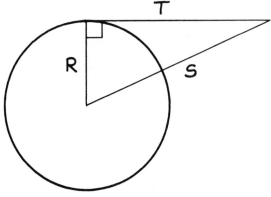

Now, if we know R and S, we can calculate the length of the tangent, T, by applying the Pythagorean theorem. This states that the square of the hypotenuse of a right triangle is equal to the sum of the squares of the other two sides. Here the formula would be: $S^2 = R^2 + T^2$.

Each player rolls the dice. The lower figure rolled represents R, while the higher figure represents S. If a player throws two equal numbers, he loses his turn.

Suppose the first player rolls a 3 and a 6. Then $6^2 = 3^2 + T^2$; $36 = 9 + T^2$; $T^2 = 36 - 9$; $T^2 = 27$; $T = \sqrt{27}$.

Using the Table of Squares and Square Roots in the Appendix, we find that $T = 5.196$, or 5.2 rounded off to one decimal place. This is the player's score.

The player who first attains a total score of 25 points is the winner.

Balloon Watching

Solving Algebraic Equations, Multiplication, Finding Tangents

Two or more players   Red and white dice

Paper and pencil

How can you find the height of a balloon—or, for that

matter, a tree or a tall building—without measuring it directly?

First you must measure the distance from your position on the ground to a spot directly under the balloon. Then, with the aid of any simple sighting device, like a telescope, you aim at the balloon and determine the angle formed between the ground line and the line between the balloon and your position on the ground.

You now have a right triangle, as shown, in which one of the angles (other than the right angle) is known (that is, angle A) as well as the side adjacent to it (d).

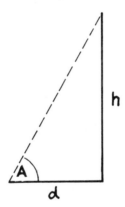

The tangent of an angle in a right triangle (other than the right angle) is the ratio between the side opposite the angle and the side adjacent to it. In other words, $\tan A = h/d$. Knowing A and d, we can calculate the height h with the aid of the table, Values of the Trigonometric Functions, in the Appendix.

Each player, in turn, rolls the dice. The figure on the red

die, multiplied by 14 (to make the figures more realistic), determines angle A. The figure on the white die, also multiplied by 14, represents the distance on the ground, in feet, from the position of the observer to a spot directly under the balloon.

Suppose that the first player rolls red 3, white 6. Then angle A is $3 \times 14 = 42°$, and distance d is $6 \times 14 = 84$ feet. Since $\tan 42° = h/84$, then $h = 84 \tan 42°$.

Consulting the table of tangents, we find that $\tan 42° = 0.9$. (In this game we use only one-place values of the tangents.)

Substituting in the equation, we have $h = 84(0.9) = 75.6 = 76$ feet (rounded off). The player's score is 76 points.

If the next player rolls red 5, white 4, angle A is $5 \times 14 = 70°$, and distance d is $4 \times 14 = 56$ feet. Then, since $\tan 70° = 2.7$, $h = 56(2.7) = 151.2 = 151$ (rounded off).

The first player to attain a total score of 500 points is the winner.

Power Struggle

Multiplication of Numbers with the Same Base But Different Exponents

Two or more players   Red and white dice

The struggle for power in this game is a contest to multiply powers of the same base by adding the exponents.

Let us recall that 2×2 may be described as "two squared" or "two raised to the second power" and represented symbolically by 2^2. Then $2^2 = 4$, and 2^3 (two cubed or two raised to the third power) $= 2 \times 2 \times 2 = 4 \times 2 = 8$. The number which is raised to a power is called the *base*, and the number representing the power is called the *exponent*. So $2^1 = 2$ to the first power; the base is 2, and the exponent is 1.

To multiply numbers with the same base but different exponents, just add the exponents: $5^2 \times 5^3 \times 5^4 = 5^9$.

In this game all numbers will be changed to base 2, 3, or 5 if they are not already powers of these bases. For example, 8^2 can be converted to base 2 by representing it as $(2^3)^2$. But this number, in turn, is $2^3 \times 2^3$. To multiply these two numbers, we add their exponents: $3 + 3 = 6$. This becomes the exponent of the product with the base 2. So $2^3 \times 2^3 = 2^6$. Another way of getting the same result is to multiply the exponents 3 and 2 in the expression $(2^3)^2$, giving $2^{3 \times 2} = 2^6$.

Each player, in turn, rolls the dice. The number on the white die represents the base; the number on the red die, the exponent. Thus a roll of white 3, red 2, means 3^2.

Suppose Player A rolls white 5, red 1. He writes the result, 5^1, on the score sheet below. If Player B (assume that there are only two players in this particular game) rolls white 3, red 3, he writes 3^3.

Now if Player A rolls white 2, red 1, he multiplies 2^1 by the score he has attained so far. Therefore his second-line entry is $2^1 \times 5^1$. If Player B rolls white 6, red 6, he must first convert 6^6 to base 2, 3, or 5. $6^6 = (2^1 \times 3^1)^6 = 2^6 \times 3^6$. The score he puts on the second line is the product of his first score, 3^3, and $2^6 \times 3^6$. $3^3 \times 2^6 \times 3^6 = 2^6 \times 3^9$.

Round No.	Player A	Player B
1	5^1	3^3
2	$2^1 \times 5^1$	$2^6 \times 3^9$
3	$2^1 \times 3^2 \times 5^1$	$2^6 \times 3^9 \times 5^3$
4	$2^{13} \times 3^2 \times 5^1$	$2^6 \times 3^9 \times 5^6$
5	$2^{17} \times 3^6 \times 5^1$	$2^{12} \times 3^9 \times 5^6$
6	$2^{23} \times 3^{12} \times 5^1$	$2^{22} \times 3^9 \times 5^6$
7	$2^{23} \times 3^{18} \times 5^1$	$2^{25} \times 3^9 \times 5^6$
8	$2^{35} \times 3^{18} \times 5^1$	

The score sheet also shows rounds 3–8. In round 3, A's roll represented 3^2; B's, 5^3. Do you see how these were multiplied in?

In round 4, A's roll represented $4^6 = (2^2)^6 = 2^{12}$. B's, 5^3 again. And so on.

Player A won in round 8 because his score was the first to have exponents adding to 50 or more.

Two-Base Toss

Converting Numbers to Equivalent Values in the Base-2 System

Two players   Timer

Two dice

The numbers we use to count things are arranged in a decimal system. This means that the exponent of 10 determines the placement of the digits in each number. $10^0 = 1$, and every one-digit number is a multiple of 10^0. Thus $3 = 10^0 + 10^0 + 10^0$. Every two-digit number in the decimal system is a multiple of 10^1 plus a multiple of 10^0. For example, 83 is $8 \times 10^1 + 3 \times 10^0 = 80 + 3$. And every three-digit number in this system is a multiple of 10^2 plus a multiple of 10^1 plus a multiple of 10^0, etc.

Any number in the decimal system can be converted to an equivalent number in another system with a different base. In this game the players convert decimal integers into binary integers, forming part of a system having 2 as the base. This particular system is of very great practical importance in modern electronics and computer technology.

The easiest way to perform the conversion is to imagine a pegboard for a game of ringtoss, as sketched. The values of the pegs, from left to right, are 8, 4, 2, and 1. These numbers can also be expressed as 2^3, 2^2, 2^1, and 2^0.

Each player, in turn, rolls a pair of dice, adds the two numbers turned up, and, within a given time limit, must convert the sum into a binary number. He can do this quickly if he imagines that he is allowed to toss no more than one ring on each peg. The binary number is formed of digits corresponding in order to the number of rings tossed on each peg.

Suppose the first player rolls a 3 and a 6. Their sum is 9. Since $9 = 8 + 1$, he imagines that he tosses a ring on 8 and another on 1. There are no (0) rings on 4 and on 2. Arranging the number of rings tossed in the order of the pegs on the board, he gets the number 1001, which is written 1001_2 to indicate that it is "to the base two."

If the second player rolls a 3 and a 4, adding to 7, he tosses one ring on the 4 peg, one on the 2 peg, and one on the 1 peg, to make his converted value 111_2. If the next roll is a 6 and a 6, adding to 12, the first player imagines himself tossing one ring on the 8 peg and one ring on the 4 peg to produce the number 1100_2.

Note that in the binary system only two digits are used: 0 and 1.

To score, count 1 point for each ring tossed. The player with the lower score wins.

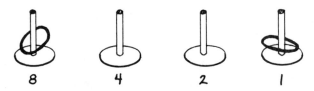

8   4   2   1

Three-Base Toss

Converting Numbers to Equivalent Values in the Base-3 System

Two players   Timer

Two dice

After a few rounds of Two-Base Toss, the players should be ready to try a trinary variation.

Use the same pegboard—but the values of the pegs, from

108

left to right, are 27, 9, 3, and 1. (Every number system, no matter what its base, gets down to 1.) These values can also be expressed as 3^3, 3^2, 3^1, and 3^0. So in this game the players convert decimal integers into base-3 numbers.

The player rolls the dice and *multiplies* the numbers turned up. Within a given time limit, he must convert the product into a base-3 number. He is allowed to toss no more than two rings on any peg. He should start with the leftmost possible peg and work toward the 1 peg.

If Player A rolls a 5 and a 3, $5 \times 3 = 15$. There are no 27's in 15, and there is only one 9, with 6 left over. Clearly the 6 can be represented by two 3's. Then the trinary number formed is 120_3. He scores $1 + 2 + 0 = 3$ points.

Notice that in the system with base 3 only three numerals are used: 0, 1, and 2.

The number 120 formed by the first player would be written as 120_3 and read as 120 "to the base 3."

If Player B rolls two 6's, whose product is 36, this is simply a 27 and a 9, giving 1100_3 for a score of 2 points.

Each player must arrive at his number within the allotted time and write it correctly. Thus, $15 = 120_3$; $36 = 1100_3$.

The winner is the player attaining the higher score after ten rounds.

Power Bases

Converting Numbers to Equivalent Values in Other Base Systems

Two or more players   Timer

Two dice

After the players have developed their skill with Two-Base Toss and Three-Base Toss, they can try this game, in which numbers to the base 10 are converted to *any* other base.

Each player, in turn, rolls the dice and sets the two numbers side by side, with the lower number to the left, to form a two-digit number. If a player rolls 4 and 5, his number is 45. Now he adds 4 and 5 to get the new base, 9. So the player who rolls 4 and 5 must convert 45 from the base 10 to the equivalent number to the base 9.

How does he convert 45_{10} to x_9?

The simplest way is to imagine that he is playing ringtoss as before. The pegs are labeled 9^3, 9^2, 9^1, and 9^0. There is nothing to toss on 9^3 or 9^2 because $9^2 = 81$, which is larger than 45. So he tries the next peg, 9^1. 45 exactly equals five imaginary rings to be tossed on this peg. So $45_{10} = 50_9$.

The player scores 5 points, which is the sum of the digits in 50.

The first player to gain a cumulative total of 50 points wins.

Ten-Base Toss

Converting Numbers from Various Base Systems to the Base-10 System

Two or more players

Three dice or playing cards

In this variation too, we imagine that rings are tossed on pegs to convert numbers from one base to another. But here we convert to the familiar base-10 system. The procedure is the reverse of that followed in Power Bases: instead of dividing, we multiply.

Each player rolls three dice, one at a time, and makes a three-digit number. If he rolls 3, 4, and 2, his number is 342. But what is the base? He adds 1 to the highest digit in the number: $4 + 1 = 5$; so his problem is to convert 342_5 to base 10. If he rolls 5, 6, and 3, his number is 563_7, that is, 563 to the base 7.

If the first player rolls 5, 3, and 4, his number is 534_6. Let's work it out. Set up a series of imaginary pegs, labeled 6^2, 6^1, and 6^0. Three pegs are enough for the three digits of 534.

Now the player multiplies each digit in his number by the value of the corresponding peg. He multiplies 5 by 6^2 to get 180 (base 10). And $3 \times 6^1 = 18$ and $4 \times 6^0 = 4$. Adding, he gets $180 + 18 + 4 = 202$. Thus $534_6 = 202_{10}$. His score is $2 + 0 + 2 = 4$ points, provided his work is shown and the answer is written correctly.

The winner is the first player to accumulate 100 points.

A more difficult variation uses a deck of playing cards instead of dice. Each player draws three cards. If he draws 9, jack (11), and 2, his number is 9, 11, 2. (Consider 11 as a single digit.) The base is 1 plus the highest digit, 11. So the base here is $11 + 1 = 12$ (the duodecimal system).

Set up the imaginary pegs, from left to right, 12^2, 12^1, 12^0. Now multiply: $9 \times 12^2 = 9 \times 144 = 1296$; $11 \times 12^1 = 11 \times 12 = 132$; $2 \times 12^0 = 2 \times 1 = 2$. Since $1296 + 132 + 2 = 1430$, the answer is 1430_{10}, and the player scores $1 + 4 + 3 + 0 = 8$.

After all the cards in the deck are used, the player with the highest score wins.

4

Games with Prepared Materials

THE MATERIALS needed for the games in this chapter can easily be prepared, often with the assistance of the players. You may need index cards, colored crayons, adhesive tabs, playing cards, coins, or dominoes. Above each game you will find a list of the necessary materials—which may also include paper and pencil, dice, a checkerboard, or a felt marking pen.

Dominations and Powers

Multiplication, Division

Two, three, or four players   Dominoes

Detachable tabs

Using detachable tabs or tapes, prepare a set of twenty-four dominoes by writing on each tab two problems, one involving division and one involving multiplication, the solutions to which equal values in the multiplication tables of 2, 3, 4, and 5. One domino in the set is illustrated.

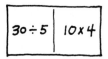

The others are labeled 18 ÷ 3; 5 × 8. 12 ÷ 2; 10 × 4. 6 ÷ 2; 18 × 2. 16 ÷ 2; 12 × 3. 30 ÷ 3; 9 × 4. 20 ÷ 2; 4 × 6. 50 ÷ 5; 7 × 4. 48 ÷ 4; 5^2. 90 ÷ 3; 4 × 3. 36 ÷ 3; 16 × 2. 64 ÷ 2; 5 × 3. 30 ÷ 2; 12 × 2. 45 ÷ 3; 5 × 5. 56 ÷ 2; 4 × 4. 48 ÷ 3; 5 × 7. 64 ÷ 2; 4^2. 70 ÷ 2; 9 × 2. 18 ÷ 2; 6 × 3. 36 ÷ 2; 14 × 2. 100 ÷ 5; 6 × 5. 80 ÷ 4; 10 × 3. 40 ÷ 2; 8 × 4. 70 ÷ 2; 8 × 3. This is set 1. Prepare other similar sets of twenty-four dominoes, using values in the multiplication tables of 6, 7, 8, and 9, and of 9, 10, 11, and 12.

Begin by playing with set 1. Turn the dominoes face down on the table and scatter them about. Each player takes five dominoes. Turn one of the dominoes remaining on the table face up.

Player 1, looking at the dominoes he has drawn, tries to match either one of the values on the face-up domino with a value on one of his own. For example, if the domino face up on the table is 80 ÷ 4; 10 × 3, as illustrated, Player 1 can attach the domino 100 ÷ 5; 6 × 5, because 6 × 5 = 10 × 3 = 30.

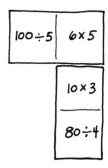

Each player, in turn, tries to attach a domino half of which has a value equivalent to that of an unmatched half on the table, as in a regular game of dominoes. Each time a player attaches a domino, he says aloud the value of the equivalent—for example, "Eight times four is thirty-two." His score is the equivalent value matched (e.g., 32 points). That is why, in the illustration, Player 1 chose to attach 6 × 5 to 10 × 3 for a score of 30 points, rather than 80 ÷ 4 to 100 ÷ 5 for a score of only 20 points. If a player makes an error in his calculation, he scores 0. If he cannot attach an equivalent value, he may take up to two dominoes from the table.

The game continues until one player has discarded all his dominoes or the game is stalled because no one can attach a domino even after drawing two extra dominoes from the table.

A player who has discarded all his pieces gets a bonus of 25 points. A cumulative total score should be kept by each player. The highest total score wins.

After a few games using set 1, go on to set 2 and ultimately set 3. The problems can be made more difficult by including squares and square roots.

Seesaw

Multiplication, Division, Fractions

Two or more players   Timer

Prepared cards

The players in this game have to perform a balancing act, taking care that the weights of the people on the seesaw are balanced so that it goes neither up nor down.

Can a 200-pound parent and a 50-pound child use the seesaw safely together? They can if they know the principle involved: the weight on one side multiplied by the distance from the pivot of the seesaw must equal the weight on the other side times its distance from the pivot. In this case, $50 \times 4 = 200 \times 1$, so that the child can be 4 feet from the fulcrum and the parent 1 foot — or 6 feet and 1 1/2 feet, or 1 meter and 25 centimeters, etc.

Prepare a set of twenty cards. (You can use 3-by-5-inch index cards.) Mark each card with a weight for the child and another for the parent. For example, one card may have: child 30, parent 120; another may have: child 40, parent 180.

Shuffle the cards and place the pile face down on the table. The players take turns drawing from the top of the pile. Within a given time limit, each player must balance parent and child on the seesaw.

Suppose that a player draws the card with child 30, parent 120. He gets 1 point for noting that 30 pounds $\times$ 4 feet = 120 pounds $\times$ 1 foot, or that 30 pounds $\times$ 2 feet = 120

pounds × 1/2 foot, or that 30 pounds × 6 feet = 120
pounds × 1 1/2 feet. 25 cumulative points win.

MADSopoly

Addition, Subtraction, Multiplication, Division, Squares

Two, three, or four players   Two dice

Prepared cards   Colored disks

Prepared board

Prepare a board as illustrated. Then prepare a thousand
dollars of play money by writing the values of the currency
on cards: five $100 bills, two $50 bills, ten $20 bills, ten
$10 bills, ten $5 bills, and fifty singles. Give each player
$200 to start with. Leave the rest on the board.

Each player starts on GO and, in turn, rolls the dice. Using
"MADS" (multiplication, addition, division, and subtraction), he combines the two numbers he rolls to form
four answers, any one of which may be used as a move.

For example, if he rolls 4 and 3, he can multiply the numbers (12), add them (7), divide the larger by the smaller
(1 1/3 = 1 rounded), or subtract the smaller from the
larger (1). Which answer should he use as a move?

Looking at the board and counting from GO at the lower
left, we note that moving twelve boxes will land him on an

117

asterisk (*); seven boxes, on $+4^2$; and one box, on $+9^2$. A plus before a number means that the player is to take from the board the amount of money indicated in the box on which he lands (16 or 81 in this case). A minus before a number indicates the amount of money he must give the board. An asterisk (*) means that he must give \$15 to the board. △ means that he is to take \$10 from each player. ○ means that he must give each player \$15. Landing on GO gives him the right to take \$50 from the board; passing GO permits him to take \$25 from the board.

Obviously, the player who rolls 4 and 3 should move one box to $+9^2$ and collect \$81 from the board. His other choices are to take only \$16, or to pay out \$15.

Differently colored disks should be used by different players. A player may not move more than twelve boxes at a time. If a player rolls 5 and 3, for example, he cannot use $5 \times 3 = 15$; but he can use $5 - 3 = 2$ or $5 \div 3 = 1$ $2/3 = 2$ rounded, or $5 + 3 = 8$. His choice depends on what square his disk is on.

-7^2	$+5^2$	-4^2	*	$+6^2$	-3^2	○	$+2^2$	-9^2	△	← GO 14
*										$+8^2$ 13
$+8^2$										* 12
○										△ 11
GO →	$+9^2$ 1	△ 2	-2^2 3	* 4	$+3^2$ 5	○ 6	$+4^2$ 7	-5^2 8	$+7^2$ 9	-6^2 10

118

The players take turns rolling the dice, advancing their disks, and receiving or paying out money, until all the money on the board is gone. Thereafter, instead of getting money from the board, the other players must contribute equally. The game continues until one player has won all the money.

Pi

Fraction Operations

Two players

Prepared cards

The Greek letter π, pronounced "pi," represents a very distinguished member of the mathematical family. It is the ratio of the circumference to the diameter of a circle and is approximately 3.1415926536, or 22/7. In this game it is the one number that is *not* the answer to any problem with fractions, and the object is to get and keep π in order to win.

Prepare fifteen cards, each with a different problem involving operations with fractions. (For the first round, confine the operations to addition and subtraction.) Prepare fifteen more cards showing the solutions. The thirty-first and last card shows π.

Here is one possible set of matching cards:

1/2 + 1/3	5/6
3/4 − 1/2	1/4
1 1/2 + 2 1/2	4
3 1/2 − 1 1/4	2 1/4
5/8 + 1/4	7/8
3/12 − 1/4	0
2/3 + 1/2	1 1/6
5 − 2 1/2	2 1/2
2 3/4 − 1 1/2	1 1/4
7/8 − 1/2	3/8
1/5 + 1/2	7/10
2 1/8 − 1/4	1 7/8
1/8 + 1/3	11/24
1/2 − 1/3	1/6
1/3 + 1/4	7/12

After adding the card marked π, shuffle the cards and deal them. (One of the players will have an extra card.)

Each player now looks at his hand and tries to find pairs of matching cards, a problem card and its answer card. The players discard the matching pairs face up so that any errors may be detected. Then the players take turns drawing one card from each other and trying to match and discard more cards.

The game continues until thirty cards have been discarded. The winner is the player left holding the card with π.

After the first round, a new set of cards can be prepared, this time also involving the multiplication and division of fractions. The problems can be made increasingly difficult in later rounds, for example, by using larger improper fractions.

120

Fraction Rummy

Fraction Operations

Two, three, or four players

Detachable labels

Deck of playing cards

Paste removable labels on the backs of ordinary playing cards. On each label write one of the following: 1/2, 1/3, 2/3, 1/4, 2/4, 3/4, 1/5, 2/5, 3/5, 4/5, 1/6, 2/6, 3/6, 4/6, 5/6, 1/8, 2/8, 3/8, 4/8, 5/8, 6/8, 7/8, 3/9, 6/9, 1/10, 2/10, 3/10, 4/10, 5/10, 6/10, 7/10, 8/10, 9/10, 1/12, 2/12, 3/12, 4/12, 5/12, 6/12, 7/12, 8/12, 9/12, 10/12, 11/12. To these forty-four add eight duplicates: 1/3, 1/4, 3/4, 1/5, 1/6, 1/8, 3/8, 1/12. Thus, the deck of prepared cards should total fifty-two.

Shuffle the cards and deal seven cards, one at a time, to each of the players. Place the rest of the deck on the table, with the top card label-side up.

Each player examines the cards in his hand and tries to add fractions to produce a sum that is a whole number (1, 2, 3, etc.). For example, 1/2 + 2/4 = 1/2 + 1/2 = 1; 3/6 + 5/12 + 1/12 = 1/2 + 6/12 = 1/2 + 1/2 = 1; 7/8 + 1/2 + 1/4 + 3/8 = 7/8 + 4/8 + 2/8 + 3/8 = 16/8 = 2. When a player finds such a sum, he places the cards on the table next to him so that the other players can check his arithmetic.

The first player begins by picking up the top card from the pack on the table and trying to make a combination. Then he discards one card label-side up. The next player can pick up the discard or take the top card from the pack, whichever will help him more to produce a combination that will add up to a whole number. Then he discards, and so on.

The game continues until one player has laid down all his cards, winning the round. If no player has won when all the cards on the table have been used, the pack may be used again.

The winner of a round scores 5 points, and each player, including the winner, scores 1 point for each card laid down as part of a combination adding up to a whole number.

When one player has cumulated 100 points he wins the match.

"SAM" Decimals

Fraction Operations

Two or more players   Deck of playing cards

Detachable tabs   Die

The game is named after the subtraction, addition, and multiplication of decimal numbers.

Prepare a die by covering the 4, 5, and 6 pips with detachable tabs marked 1, 2, and 3. Now the die has two 1's, two 2's, and two 3's.

Separate the deck of cards into two packs, one with the red suits (diamonds and hearts) and one with the black suits (clubs and spades). The cards of the black suits are treated as whole numbers and the cards of the red suits are considered as decimals. Thus if a player draws a black 6 and a red jack (.11), his number is 6.11. In this game the jack counts as 11 (or .11); the queen, 12 (or .12); and the king, 13 (or .13).

Player 1 rolls the die. If he throws 1, he draws one card from each pack and multiplies the two values he thus draws. If he throws 2, he draws twice (two cards from each pack), and subtracts the smaller value from the larger. If he throws 3, he draws three cards from each pack and adds the values of the three pairs.

If a player makes an error in subtraction, addition, or multiplication, he scores 0. Scores are kept cumulatively by each player. The winner is the first player to score 50 or more points.

Let us follow a typical game with two players.

Say that Player 1 throws a 2. He draws black 3, red queen (3.12); and black 10, red 7 (10.7). Subtracting the smaller number from the larger, he scores $10.7 - 3.12 = 7.58$, rounded to 8. (Or the players can round to one decimal, which would be 7.6.)

Player 2 throws a 1 and draws black jack, red 3 (11 and 0.3). Multiplying, he scores $11 \times (0.3) = 3.3$, or 3. The score is 8 to 3, with Player 1 leading.

In the second round, Player 1 throws a 3. He draws black 4, red 1 (4.1); black 8, red 5 (8.5); and black 6, red 2 (6.2).

Adding, he scores $4.1 + 8.5 + 6.2 = 18.8$, or 19 points. Player 2 throws a 1. He draws black king, red 5 (13 and 0.5). Multiplying, he scores $13 \times (0.5) = 6.5$, or 7 (rounding up). The score is now 27 $(8 + 19)$ to 10 $(3 + 7)$ in favor of Player 1; so play continues.

Player 1 next rolls a 2. He draws black 1, red 2 (1.2) and black jack, red king (11.13). Subtracting the smaller from the larger number, he scores $11.13 - 1.20 = 9.93$, or 10 points. Player 2 rolls a 2 also. He draws black 2, red 9 (2.9) and black 9, red 8 (9.8). Subtracting the smaller from the larger value, he scores $9.8 - 2.9 = 6.9$, or 7 points. Now Player 1 has a cumulative score of 37 $(27 + 10)$, and Player 2 has a score of 17 $(10 + 7)$.

On his next roll Player 1 turns up a 3. He draws black 7, red 11 (7.11); black 8, red 12 (8.12); and black 3, red 8 (3.8). Adding, he scores $7.11 + 8.12 + 3.80 = 19.03$, or 19 points. This, added to his score of 37, brings his cumulative score to 56. So Player 1 wins.

Several rounds should be played.

Arithmessages

Addition, Subtraction, Multiplication, Division, Fractions

Two players   Timer

Prepared cards

Children love to be privy to secrets. They delight in using private languages for messages that nobody but the initiated can understand. By the same token, their curiosity is aroused instantly by a rebus, a riddle, or a cryptogram. Intrigued by the mystery, they will spend hours trying to crack a cipher.

Encoding or decoding the ciphers in this game and its variations that follow has a double educational value. Because of the transition from words to numbers, every game in this series of mathematical ciphers and cryptograms combines practice in such verbal skills as spelling, vocabulary usage, and sentence construction with basic arithmetical operations.

To send arithmetical messages, you need a letter-number cipher like the one shown here. A quick way to recall the correlations between letters and numbers is to place them on three cards, as illustrated.

A$_1$	B$_2$	C$_3$
D$_4$	E$_5$	F$_6$
G$_7$	H$_8$	I$_9$

J$_{10}$	K$_{11}$	L$_{12}$
M$_{13}$	N$_{14}$	O$_{15}$
P$_{16}$	Q$_{17}$	R$_{18}$

S$_{19}$	T$_{20}$	U$_{21}$
V$_{22}$	W$_{23}$	X$_{24}$
Y$_{25}$	Z$_{26}$	

The key can be kept in mind by remembering that $J = 10$ and $S = 19$.

Using this cipher, you can write "Joe" as 10, 15, 5, and "Mary" as 3, 1, 18, 25.

Within an agreed time limit, each player makes up a secret message based on the cipher, but the numbers are further disguised as mathematical problems. For example, 13 + 6, 3 + 2, 8 − 3, means 19, 5, 5, which in turn designates the word "see." The combination 5 × 5, 9 + 6, 35 − 14, signifies 25, 15, 21, or "you." Similarly, 4 × 3, 13 − 12, 12 + 8, 20/4, 9 × 2, denotes 12, 1, 20, 5, 18, or "later"; and 6 + 4, 8 + 7, 9 − 1, means 10, 15, 5, or "Joe." So the whole set means "See you later, Joe."

When time is up, the players exchange messages and try to decode them in the same amount of time they were given to write them. A player scores 2 points for each expression he deciphers correctly and 1 point more for each finished word. In the message above, there are fourteen expressions (28 points) and four words (4 points). The score for deciphering the message is therefore 32 points.

Of course, the mathematical problems can be made more baffling: 39/3 = 13 = M; 13 1/2 − 4 3/4 = 8 3/4, or 9 (rounded off) = I; (5 × 8) − 16 = 40 − 16 = 24 = X; 3418 − 3397 = 21 = U; (2/3 of 27) − 2 = 18 − 2 = 16 = P. And that is how you write "mix-up" in our arithmetical cipher!

Greeting Cards

Addition, Subtraction, Multiplication, Division, Fractions, Square Roots, Algebraic Equations

Two or more players   Prepared cards

Paper and pencil   Timer

In this more challenging variation the messages sent by

each player are greetings enciphered in mathematical language.

Each player makes up a greeting to be sent on a greeting card to another player. He then translates his greeting into numbers, using the letter-number cipher in Arithmessages. Each number is represented on the greeting card by a mathematical problem, which is made as difficult as possible. The players exchange greeting cards and, within an agreed time limit, proceed to decode them.

The first player to decode his greeting card gets 5 points, plus (like the other players) 1 point for each problem solved and 1 point for each word deciphered correctly.

Thus, if a player correctly decodes a greeting card with the message 2^3, $3^2 - 8$, 4^2, $64 \div 4$, 5^2, $\sqrt{16} - 2$, 3^2, $27 \times 2/3$, $2^2 + 16$, $24/3$, 2^2, $5^1 - 4$, $2x - 5 = 45$, he scores 13 points for correctly solving the thirteen problems, 2 points for the two words "Happy Birthday," and, if he is the first to decode the greeting, the bonus 5 points, for a total of 20 points.

The winner is the first player to accumulate 100 points.

Numbered Names

Addition, Subtraction, Multiplication, Division, Fractions, Finding Square Roots, Solving Algebraic Equations

Two or more players   Prepared cards

Paper and pencil   Timer

This variation uses the letter-number values of Arithmessages for coding and decoding names — not just the names of ordinary people, but also names chosen from history, literature, or geography, to review any subject matter agreed on by the players and suitable to their interests and maturity. Thus, CAESAR, HANNIBAL, and ROOSEVELT may be among the names used for several rounds in history; SHAKESPEARE, MILTON, and HOMER could figure in a set on literature; and WISCONSIN, TEXAS, and WYOMING would be suitable for a round with the names of the states. Other rounds can be played with the names of birds, animals, games, tools, kitchen utensils, etc.

Each player makes up a number of mathematical problems that, when their solutions are deciphered, do not spell the name but an anagram of it — say, CARSEA for CAESAR. The players then exchange the cards and proceed to decipher them. A player scores 1 point for each problem solved correctly within the agreed time limit, 1 point for rearranging the letters to form the name, and 5 points if he is the first one to finish.

The problems should be as difficult as possible and can involve square roots, algebraic equations, fractions, and all the operations of arithmetic.

Alphamerics

Addition, Subtraction, Multiplication, Division, Fractions, Decimals

Two players only   Prepared cards

Paper and pencil   Timer

Have you ever added words? It can be done, you know.

For example, RAIN + WIND = STORM.

This is, of course, true. But can you prove it mathematically?

You can do so if you use numbers to show that it is correct.

Here are some hints. Let $A = 3$; $N = 2 \times A$; $D = (N/2) + 1$; $W = 2 \times D$; and $I = N/3$. This means that $N = 2 \times 3 = 6$; $D = 6/2 + 1 = 3 + 1 = 4$; $W = 2 \times 4 = 8$; and $I = 6/3 = 2$. Remembering that each letter always stands for the same digit, we can determine the numerical values of the other letters: S, T, O, R, and M. Substituting the numerical values for the letters we already know, we prove the correctness of the sum in this way:

RAIN	R326	R326	9326
WIND	8264	8264	8264
STORM	STORM	ST590	17590

Each player first works out, within a given time, a word sum like the one above and provides a few clues to the numerical values of the letters. These hints can be made as complicated as desired and may involve problems with fractions and decimals and all the operations of arithmetic. When time is up, the players exchange the cards containing their word sums and, within an equal time limit, try to prove the correctness of the addition by finding the numerical values of all the letters.

A player earns 2 points for each correct letter value that he finds. So in the alphameric shown above a player would score 2 points for each of the five letters of STORM, or 10 points in all.

The winner is the player with the higher score after five rounds.

Here are a few alphamerics, with their complete solutions (but without clues):

ONE	621	TWO	935
TWO	846	SEVEN	58682
FIVE	9071	ELEVEN	878682
EIGHT	10538	TWENTY	938299

FOOD	9551
+ FAD	+ 931
DIETS	10482

SEND ÷ A = GIFT
7852 ÷ 4 = 1963

BASE	7483
+ BALL	+ 7455
GAMES	14938

And here are a few more alphamerics with clues and the solutions based on them:

FOOT + BALL = GAME. $F = 2$; $T = (3 \times F) + 1$; $B = F^2$; $A = F + 1$; $L = T + 1$. Find G, M, E, and O.

$T = (3 \times 2) + 1 = 6 + 1 = 7$; $B = 2^2 = 4$; $A = 2 + 1 = 3$; $L = 7 + 1 = 8$. In two steps:

FOOT	2OO7	2OO7
+ BALL	+ 4388	+ 4388
GAME	G3ME	6395

MOON + MEN + CAN = REACH. $O = 5$; $M = O + 4$; $C = M/3$; $A = M - 1$; $H = 2 \times C$. Find E, R, and N.

$M = 5 + 4 = 9$; $C = 9/3 = 3$; $A = 8$; $H = 6$. In three steps:

130

MOON	955N	955N	9552
MEN	9EN	90N	902
CAN	38N	38N	382
REACH	RE836	10836	10836

FISH + CHIP = MEAL. I = 1; H = $(7 \times I) - 4$; S = H^2; C = 2H; M = S − 1; L = I + 6. Find F, E, A, and P.

H = $(7 \times 1) - 4 = 7 - 4 = 3$. S = $3^2 = 9$. C = $2 \times 3 = 6$. M = 9 − 1 = 8. L = 1 + 6 = 7. In two steps:

FISH	F193	2193
+ CHIP	+ 631P	+ 6314
MEAL	8EA7	8507

A famous alphameric was used by one conspirator to send a message to another. The message was a sum:

FLY
FOR
YOUR
LIFE

The sum can be deciphered with the clues that I = 1; U = E^2; F = R − 2; L = U + 5; and O = 0. Then F must be 5, and so on. The answer is:

598
507
8047
9152

Here are two more interesting alphamerics with their clues.

HOT
+ COLD
WARM

131

Find W, A, R, and D.

$H = 2$; $O = (4H) - 1$; $T = H^2$; $C = H \times T$; $L = 36 \div (T + 2)$; $M = T + 1$.

Solution: $W = 9$; $A = 0$; $R = 3$; $D = 1$.

$$\begin{array}{r} BANK \\ + \underline{NOTE} \\ CASH \end{array}$$

$N = 1$; $K = 2N$; $A = K^2$; $E = (5N) - 2$; $T = E + N + K$; $B = 7 + N$.

With a little practice, anyone can produce alphamerics and solve them if given a few clues.

Alphabetical Figures

Solving Algebraic Equations

Two or more players   Two dice

Prepared cards   Timer

In algebra, letters stand for numbers. In this game figures are substituted for letters to solve equations.

First, prepare twenty or more cards, depending on the number of players, each with a different notation, like the following: $l + s$, $2l - s$, $l^2 + s$, $l^2 + s^2$, $l^3 - s$, ls, $l + 3s^2$, $l^2 - s^2$, $l^3 + 2s$, $l + 3s^2$, $(l + s)^2$, $(l - s)^3$, $l^2 + 2ls + s^2$, $\dfrac{2l + s}{2}$, $l + s^3$, $l^3 - s^3$, $(l + 2s)^2$, $5l - 2s$, $(l + 3s)^2$.

132

Shuffle the cards thoroughly and place them face down on the table.

The first player rolls the dice. Say he throws 5 and 4. He picks a card from the top of the pile, say, the card marked $2l - s$. In this game the letter l always stands for the larger number on the dice, and s for the smaller. (If the numbers are equal, as in 5, 5, then $l = s$.) Here the larger number is 5, and the smaller is 4. So $2l - s = 2(5) - 4 = 10 - 4 = 6$. If the player can solve the equation within the time limit, he scores its value: in this case, 6 points.

If, for example, the next player rolls 3, 6, and picks a card showing $l^2 - s$, his score is $6^2 - 3 = 36 - 3 = 33$ if he does the calculation correctly and in time.

When one player has accumulated 1000 points or more he wins. If the cards are used up before this happens, the winner is the player with the higher or highest score.

In other rounds different formulas, varying in difficulty according to the players' abilities, can be written on the cards.

Fraction Locus Bingo

Addition of Fractions

Two or more players   Two dice

Prepared cards   Timer

Each player prepares his scorecard by writing on it his

own arrangement, in a different order, of the fractions in the grid shown. Note the free space in the center.

$\frac{1}{2}$	$\frac{5}{6}$	$\frac{13}{12}$	$\frac{3}{4}$	$\frac{22}{15}$	$\frac{21}{20}$	$\frac{37}{30}$
$\frac{7}{12}$	$\frac{31}{20}$	$\frac{2}{3}$	$\frac{17}{15}$	$\frac{2}{5}$	$\frac{4}{3}$	$\frac{43}{30}$
$\frac{23}{30}$	$\frac{1}{3}$	$\frac{9}{20}$	$\frac{11}{12}$	$\frac{13}{10}$	$\frac{13}{20}$	$\frac{29}{30}$
$\frac{7}{10}$	$\frac{8}{15}$	$\frac{3}{5}$	FREE	$\frac{39}{30}$	$\frac{49}{30}$	$\frac{17}{20}$
$\frac{14}{15}$	$\frac{19}{15}$	$\frac{13}{15}$	$\frac{9}{10}$	$\frac{16}{15}$	$\frac{11}{15}$	$\frac{17}{30}$
$\frac{11}{10}$	$\frac{13}{10}$	$\frac{7}{6}$	$\frac{5}{3}$	$\frac{19}{12}$	$\frac{21}{20}$	$\frac{4}{5}$
$\frac{3}{2}$	$\frac{19}{20}$	$\frac{5}{4}$	$\frac{27}{20}$	$\frac{11}{30}$	$\frac{17}{12}$	$\frac{31}{30}$

The leader rolls the dice twice and calls out the numbers — say, 1, 3, and 2, 5.

Each player then silently forms the fractions 1/3 and 2/5, using the smaller number in each pair as the numerator and the larger as the denominator. He adds the two fractions (11/15) and, within a given time limit, checks off the sum on his grid. The box containing this fraction has been checked on the illustrated scorecard.

The leader keeps a record of the fractions that were added and checks their sum on his own grid for later verification of players' scorecards.

When the fractions add up to whole numbers, the dice must be rolled again, because whole numbers do not appear on the grid. Thus, if the numbers turned up are 2,2, the dice must be rolled again. Similarly, if the two pairs of numbers are, for example, 4,6 and 2,3, as their sum is 6/6 or 3/3, a whole number, another roll of the dice is required.

The game continues until a player has checked three contiguous boxes on his grid. He then yells out, "Fraction Bingo!" and has his work checked by the leader. If all is well, he is the winner of the round.

Note that a variety of winning patterns is possible, such as:

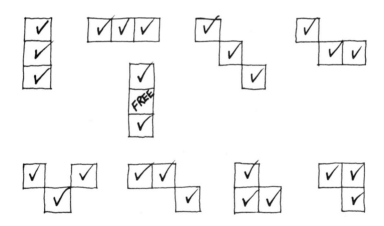

All fractions should be reduced before they are added. For example, 2/4 and 3/5 should be added as 1/2 and 3/5, to get the sum 11/10.

Product Bingo

Multiplication of Fractions

Two or more players   Prepared cards

Paper and pencil   Timer

In this variation of Fraction Locus Bingo, a set of sixteen cards is prepared, each with one of the following problems in the multiplication of two fractions: $(1/2 \times 1/3)$, $(1/2 \times 3/4)$, $(5/6 \times 1/2)$, $(7/8 \times 1/2)$, $(2/3 \times 1/3)$, $(3/4 \times 2/3)$, $(2/3 \times 5/6)$, $(7/8 \times 2/3)$, $(4/5 \times 1/3)$, $(4/5 \times 3/4)$, $(5/6 \times 4/5)$, $(4/5 \times 7/8)$, $(1/6 \times 7/8)$, $(1/3 \times 1/6)$, $(3/4 \times 1/6)$, $(5/6 \times 1/6)$.

Each player prepares a scorecard by writing on it his own arrangement, in a different order, of the fractions shown:

$\frac{1}{6}$	$\frac{2}{9}$	$\frac{4}{15}$	$\frac{1}{18}$
$\frac{3}{5}$	$\frac{1}{2}$	$\frac{3}{5}$	$\frac{1}{8}$
$\frac{5}{12}$	$\frac{5}{9}$	$\frac{2}{3}$	$\frac{5}{36}$
$\frac{7}{16}$	$\frac{7}{12}$	$\frac{7}{18}$	$\frac{7}{48}$

The leader—parent or teacher—shuffles the multiplication cards and places them face down on the table. He then draws the top card and reads the problem. Within a given time limit, each player multiplies the two fractions and checks off the product on his grid. The leader does the same on his grid for later verification of the players' scorecards.

The game continues until a player has checked off four boxes forming a vertical column, a horizontal row, or a diagonal line. He then yells, "Product Bingo!" and has his work checked by the leader. If it is found to be correct, he wins the round.

Quotient Bingo

Division of Fractions

Two or more players   Prepared cards

Paper and pencil   Timer

This variation of Fraction Locus Bingo is played like Product Bingo, except that the sixteen prepared cards show division of fractions: $(1/6 \div 1/3)$, $(1/6 \div 3/4)$, $(1/6 \div 5/6)$, $(1/6 \div 7/8)$, $(2/3 \div 1/3)$, $(2/3 \div 5/6)$, $(2/3 \div 3/4)$, $(2/3 \div 7/8)$, $(4/5 \div 1/3)$, $(4/5 \div 3/4)$, $(4/5 \div 5/6)$, $(4/5 \div 7/8)$, $(1/2 \div 1/3)$, $(1/2 \div 3/4)$, $(1/2 \div 5/6)$, $(1/2 \div 7/8)$.

Each player prepares a scorecard by writing on it his own arrangement, in a different order, of the fractions shown:

$\frac{1}{2}$	2	$\frac{12}{15}$	$\frac{3}{2}$
$\frac{2}{9}$	$\frac{8}{9}$	$\frac{16}{15}$	$\frac{2}{3}$
$\frac{1}{5}$	$\frac{4}{5}$	$\frac{24}{25}$	$\frac{3}{5}$
$\frac{4}{21}$	$\frac{16}{21}$	$\frac{32}{35}$	$\frac{4}{7}$

In all other respects the game is played like Product Bingo.

137

Binomial Bingo

Multiplication of Binomials

Three or more players   Dice

Prepared cards   Disks

A quick way to multiply two binomials is to use the mnemonic FLIO, where F stands for the product of the two FIRST terms, L stands for the product of the two LAST terms, I stands for the product of the two INSIDE expressions, and O stands for the product of the two OUTSIDE expressions. Add them, simplify, and rearrange.

Let us apply this procedure to $(x + 5)(x + 4)$. The product of the first terms is x^2. The product of the last terms is 20. The product of the two inside terms is $5x$. And the product of the two outside terms is $4x$. $x^2 + 20 + 5x + 4x = x^2 + 9x + 20 = (x + 5)(x + 4)$.

Each player receives a scorecard with the products of various binomials written in boxes, as illustrated for Players A, B, and C. A record of dice rolls is also illustrated.

Player A rolls the dice. The two numbers rolled are the numbers to be added to x in the binomials. If he rolls 3 and 4, the binomials to be multiplied are $x + 3$ and $x + 4$. $(x + 3)(x + 4) = x^2 + 7x + 12$. It is Player B who has this product on his scorecard. If he recognizes it as the correct product of the binomials, he places a disk over the box containing it. At the same time a record of the dice roll is made as shown.

138

Now Player B throws the dice, then Player C, and so on. The winner of the round is the player who first covers a row, a column, or a diagonal on his scorecard with disks. Each round scores 5 points. Several rounds should be played, with different cards containing different products and different arrangements. The winner is the player who first accumulates 20 points.

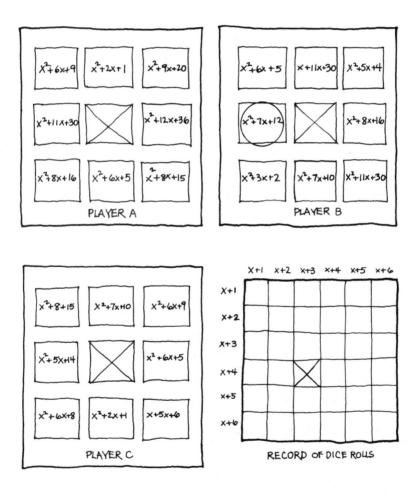

x^2+6x+9	x^2+2x+1	$x^2+9x+20$
$x^2+11x+30$	✕	$x^2+12x+36$
$x^2+8x+16$	x^2+6x+5	$x^2+8x+15$

PLAYER A

x^2+6x+5	$x+11x+30$	x^2+5x+4
$x^2+7x+12$	✕	$x^2+8x+16$
x^2+3x+2	$x^2+7x+10$	$x^2+11x+30$

PLAYER B

x^2+8+15	$x^2+7x+10$	x^2+6x+9
x^2+5x+4	✕	x^2+6x+5
x^2+6x+8	x^2+2x+1	$x+5x+6$

PLAYER C

	X+1	X+2	X+3	X+4	X+5	X+6
X+1						
X+2						
X+3						
X+4			✕			
X+5						
X+6						

RECORD OF DICE ROLLS

139

Bingo Algebra

Multiplication of Binomials

Three or more players	Red and white dice
Prepared cards	Disks

This is a slightly more difficult variation of Binomial Bingo. It is played in the same way, but both positive and negative numbers are used.

The numbers shown on the white die are positive; those shown on the red die are negative. So if a player rolls white 3, red 5, the binomials are $x + 3$ and $x - 5$. Now $(x + 3)(x - 5) = x^2 - 2x - 15$. This product does not appear on any of the cards for Players A, B, and C. But a roll of white 5, red 5, would produce $(x + 5)(x - 5) = x^2 - 25$, which appears on the scorecard of Player B, as illustrated.

Scoring is the same as in Binomial Bingo.

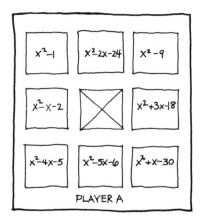

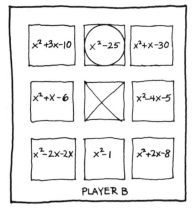

140

All possible combinations are:

$x^2 - 1$	$x^2 + 2x - 8$	$x^2 - 3x - 4$	$x^2 - x - 20$
$x^2 + x - 2$	$x^2 + 3x - 10$	$x^2 - 2x - 8$	$x^2 - 25$
$x^2 + 2x - 3$	$x^2 + 4x - 12$	$x^2 - x - 12$	$x^2 + x - 30$
$x^2 + 3x - 4$	$x^2 - 2x - 3$	$x^2 - 16$	$x^2 - 5x - 6$
$x^2 + 4x - 5$	$x^2 - x - 6$	$x^2 + x - 20$	$x^2 - 4x - 12$
$x^2 + 5x - 6$	$x^2 - 9$	$x^2 + 2x - 24$	$x^2 - 3x - 18$
$x^2 - x - 2$	$x^2 + x - 12$	$x^2 - 4x - 5$	$x^2 - 2x - 24$
$x^2 - 4$	$x^2 + 3x - 18$	$x^2 - 3x - 10$	$x^2 - x - 30$
$x^2 + x - 6$	$x^2 + 2x - 15$	$x^2 - 2x - 15$	$x^2 - 36$

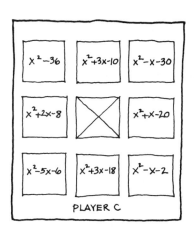

PLAYER C

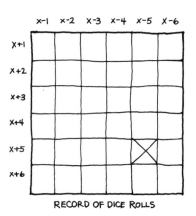

RECORD OF DICE ROLLS

Mathematical Chess

Addition, Subtraction, Multiplication, Division, Fraction and Decimal Operations

Two or more players

Paper and pencil

Prepared cards

Colored disks

First prepare a grid, as illustrated.

1 ROOK	2 KNIGHT	3 BISHOP	4 BISHOP	5 KNIGHT	6 ROOK
x2	+6	-3	+6	x3	-2
+2	SAFE +10	+8	x4	SQUARE	x4
-5	+2	+4	CUBE	-6	+7
x7	SQUARE	+11	x5	SAFE +10	+6
x5	-9	SAFE +10	+3	-9	-2
+4	+12	x6	-5	x3	+8

Each player receives three disks of the same color, one marked B to represent his bishop, one marked R to represent his rook, and one marked N to represent his knight. (If there are only two players they can use real chess pieces, one player taking white and the other black.) If more than two play, each player should have distinctively colored disks.

Prepare twelve cards marked as follows: B6, B8, B9, B12, R5, R7, R10, R9, N7, N9, N11, N13. Shuffle the cards and place them face down on the table.

As in chess, a bishop moves one or more squares diagonally, a rook moves one or more squares horizontally or vertically, and a knight moves either two squares vertically and one horizontally or two horizontally and one vertically.

The first player draws a card from the top of the pile. Suppose it is B8. Since the B on his card stands for bishop, he uses his bishop or his bishop disk. Columns 3 and 4 on the grid are marked "Bishop." Starting from the top of either column, he may move his bishop diagonally, to the left or to the right, as many squares as he wishes, to get the highest possible score. If, for example, he moves diagonally leftward from the top of column 4 to the square marked "+8," the 8 on his card is added to the 8 in the square for a score of 16. But if he chooses to move diagonally rightward from the top of column 4 to the box marked "Square," the 8 on his card is squared, for a score of 64. So this is his best possible move. He leaves his bishop or bishop disk on this box.

Suppose the next player draws N9. He would like to capture the first player's bishop. But this is impossible, because the knight cannot move to that square, whether it starts from the top of the second column or the top of the fifth column. But starting from the top of the fifth column and moving two squares vertically and one horizontally leftward, his knight can land on the square marked "Cube." This makes it possible for him to cube the number on his card, 9, for a big score of 729. He leaves his knight on this box.

If a disk is captured, the capturer scores 25 points, and his disk remains on the square of the captured piece. If a player loses all his pieces, he is eliminated from the game. If he loses a piece and draws a card requiring a move with that piece, he loses his turn. But if a disk lands on a square marked "Safe" it cannot be captured, and the player scores 10 points.

The game continues until one player has scored 1000 points. He wins.

A more complicated variation of this game makes use of a grid containing fractions and decimals.

Paired Figures

Solving Simultaneous Equations

Two or three players   Paper and pencil

Prepared cards

From two simultaneous equations with two unknowns you can find both unknown numbers. You first solve one equation for an unknown, and then substitute the value found for that unknown in the second equation.

Prepare three sets of five cards each, as follows:

Set I: S3D1, S8D2, S10D4, S9D5, S15D3.
Set II: S13P40, P56D1, S13P36, D4P96, D1P20.
Set III: S12Q5, D5Q4/3, P36Q1, P45Q5, P36Q4.

On these cards S stands for the sum of two unknown numbers, D for the difference between the larger and the smaller, P for their product, and Q for their quotient. Thus, S3D1 refers to two numbers whose sum is 3 and whose difference is 1. To find these two unknowns, set up two simultaneous equations, using l to stand for the larger number and s for the smaller:

$$l + s = 3$$
$$\underline{l - s = 1}$$
$$2l = 4$$
$$l = 4/2 = 2$$

(In this case it was faster to add the two equations, term by term, than to substitute.) Substituting in the first equation the value 2 for l, we get:

$$2 + s = 3$$
$$s = 3 - 2 = 1$$
$$l = 2, \ s = 3$$

Check: $2 + 1 = 3, \ 2 - 1 = 1$.

Shuffle the cards, mixing the three sets together. The first player draws a card and solves the problem on it by setting up simultaneous equations. If his solution is correct, he receives 5 points for a problem in set I, 10 points for a problem in set II, and 15 points for a problem in set III.

Let us solve another problem, this time from set III: P36Q4.

Again we use two simultaneous equations:

$$l \times s = 36$$
$$l/s = 4$$

Solving the second equation for l, we get:

$$l = 4s$$

Substituting this value of l in the first equation:

$$4s(s) = 4s^2 = 36$$
$$s^2 = 36/4 = 9$$
$$s = 3; \ s = -3$$
$$l = 4(3) = 12; \ l = 4(-3) = -12$$

Check: $36 = (12)(3) = (-12)(-3)$; $4 = 12/3 = (-12)/(-3)$.

Note the two sets of answers.

A player who solved this problem would earn 15 points.

The winner is the player who accumulates 100 points first.

The difficulty of the game can be adjusted by changing the numbers used on the prepared cards.

Paired Points

Solving Simultaneous Equations, Plotting Points

Two or more players   Detachable labels

Pencil   Two decks of playing cards

Graph paper   Colored disks

Prepare a square grid of one hundred boxes, with ten vertical columns and ten horizontal rows. Number the columns and rows as illustrated, labeling the horizontal axis x and the vertical axis y.

Then, on detachable labels to be pasted on the backs of two decks of playing cards, prepare one hundred sets of two simultaneous equations each, in x and y, such that every value of x from 1 through 10 and every value of y from 1 through 10 are represented in combination—for example, x 10, y 1; x 1, y 10; x 10, y 2; x 2, y 10; x 10, y 3; x 3, y 10; . . . to x 10, y 10. The pairs of equations can involve all the operations of arithmetic.

Here are a few examples of such sets: $x + y = 6$; $-x + y = 4$. $x - y = 6$; $x + y = 14$. $x + y = 10$; $-x + y = 8$. $x - y = 0$; $2x + y = 27$. $x - y = 2$; $x + y = 14$. $x + 3y = 10$; $x + y = 8$. $x - y = 3$; $x + y = 7$. $x - y = 3$; $x + y = 9$. $y - x = 2$; $y + 2x = 20$. $x + 5y = 20$; $x = 5y$. $x + y = 10$; $2x + y = 13$.

Shuffle the cards and lay them on the table with the equations face down.

Each player, in turn, draws a card from the top of the deck. After turning it up so that all can see it, he solves the equations for x and y and places a distinctively colored disk in the box on the graph whose position on the x- and y-axes corresponds to the values of x and y.

Suppose a player draws a card with the equations $x/2 + 3y = 34$; $y = 5x/4$. Substituting in the first equation the value of y given in the second:

$$x/2 + \frac{3(5x)}{4} = \frac{2x}{4} + \frac{15x}{4} = \frac{17x}{4} = 34$$

$$17x = (4)(34) = 136$$
$$x = 136/17 = 8$$

Substituting 8 for x in the second equation:

$$y = \frac{(5)(8)}{4} = 40/4 = 10$$

Testing by substituting in both equations:

$$\frac{8}{2} + (3)(10) = 4 + 30 = 34; \quad 10 = 5 \times 8 \div 4.$$

147

The player, after solving this problem, places one of his disks in the box for $x = 8$, $y = 10$, as illustrated.

If a player finds that the box is already occupied by the disk of another player (unlikely if all the cards have different combinations of x and y values), he may place his disk on top of the one he finds there.

The goal for winning can be decided by the players in advance. It can be getting three disks in a straight line or forming a square with four disks placed in adjacent boxes. The first to reach the goal is the winner.

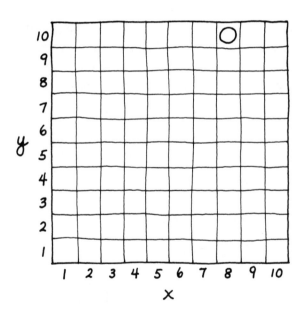

Magic Numbers

Extracting Roots, Raising Numbers to Exponential Powers, Solving Algebraic Equations, Constructing Magic Squares

Two or more players   Prepared cards

Colored pencils

This is a variation of Magic Squares, page 25. Following the procedure described, each player prepares two or three magic squares of nine boxes each. The sum of the numbers in any row, column, or diagonal is the same, the *magic constant.*

Each player then converts the numbers in his magic squares to equivalent expressions or equations. Thus, the elementary magic square shown could be converted as illustrated.

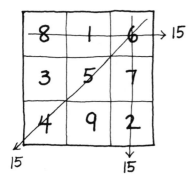

Next, each player prepares a set of ten cards. On each of nine of these he writes one of the numbers in one of his original magic squares. On the tenth card he writes a number that does not belong in that square. Thus, a player using the magic square shown above would prepare cards numbered 1, 2, 3, 4, 5, 6, 7, 8, 9, and 10, the last being unrepresented in the square. As we have seen, from an elementary magic square we can make many other magic squares by choosing one number and adding, subtracting, multiplying, or dividing that number into each box. Accordingly, if all the numbers in the simple magic square are multiplied by 2, the numbers on the corresponding cards would be 2, 4, 6, 8, 10, 12, 14, 16, and 18, and the tenth card could carry 20 or any other number. The magic constant of the elementary square is 15; that of the doubled square is 30. The player could, instead, divide each number in the original square by 3, add 5 to it, or subtract 8 from it.

Player 1 now shows one of his algebraic squares to the other players, shuffles the corresponding ten number cards, and places them face down on the table. Picking one of them up to start the game — say, 8 — he consults his original magic square to find the box that has an algebraic expression equivalent to 8. In this case, it is 2^3. He checks the box with this expression, using a pencil of a distinctive color — for example, red.

Player 2 then turns up one of the nine remaining cards and follows the same procedure, marking the appropriate square with a different color — say, blue.

The game continues until all Player 1's cards have been drawn and his square has been fully marked. Then Player 2 submits one of his magic squares and the corresponding set of number cards for the second round.

As an expression is checked in the magic square, the player who has calculated its value scores the equivalent. Thus, a player who checks an expression equalling 9 scores 9 points. When a player has completed any row, column, or diagonal, whether or not all the checks in the line are his, he scores the magic constant of the square in addition to the value of the number he checks. And if a player draws the number that does not correspond to the value of any expression in the square, and identifies it as such, he scores a bonus equal to twice the magic constant.

The winning score can be set at 100, 200, or 500, depending on the numbers used in the magic squares; and the game can be made more challenging by using negative numbers, fractions, and decimals.

Trying Triangles

Applying the Pythagorean Theorem

Two players   Timer

Prepared cards

Pythagoras proved that the square of the hypotenuse of a right-angled triangle is equal to the sum of the squares of the other two sides. This means that if one side is, for example, 3 inches, feet, centimeters, or the like, and the other side is 4, then the hypotenuse must be $\sqrt{3^2 + 4^2} = \sqrt{9 + 16} = \sqrt{25} = 5$. This ratio of 3:4:5 can be extended to multiples; say, 6, 8, and 10. Thus, $6^2 + 8^2 = 10^2 = 36 +$

$64 = 100$. Other Pythagorean ratios include $5:12:13$ and $8:15:17$.

This information can be used in quickly calculating the length of one side of a right-angled triangle if the lengths of the other two sides are known and are related to each other by any of these ratios.

Prepare fifteen cards, each showing a right-angled triangle, but with a different combination of two known sides and one unknown side. On each of five cards indicate a card value of 5 points and use one of the five following combinations of lengths for the two known sides, indicating the unknown side by x, as shown in the illustration: $6:8:x$;

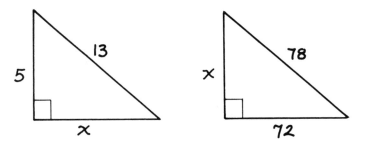

$5:x:13$; $x:15:17$; $9:x:15$; and $30:x:50$. The next five cards should each show a value of 10 points, because the calculations are somewhat more difficult. Each should contain one of these ratios: $27:x:45$; $x:24:26$; $36:48:x$; $24:x:51$; and $x:72:78$. The last five cards should be valued at 15 points each because they contain the most difficult combinations: $50:120:x$; $39:52:x$; $40:x:85$; $45:x:117$; and $80:150:x$.

The cards are now shuffled and laid face down in a pack.

Each player, in turn, draws a card from the top of the pack and, within a given time limit, must find the length of the unknown side. If his calculation is completed correctly within the given time, he scores the value shown on the card. If not, he scores 0, and his opponent tries to solve the problem, scoring the value shown on the card if he solves it in time and drawing again.

The player who first accumulates 50 points is the winner.

5
Party Games

THE GAMES in this chapter are suitable for groups of six or more players. Some are team games. They are best played in large families, at social gatherings and parties, in the classroom or at a school assembly, or among campers.

Handy Numbers

Addition, Subtraction, Multiplication, Division

Six or more players, divided into two teams

Timer

The object of the game is to interpret a series of hand

154

signals for various mathematical operations and to solve the resulting problems, all within a given time limit.

Crossed arms mean multiplication; hands held up mean addition; palms laid crosswise on each other mean division; and hands placed sidewise across the front of the neck mean subtraction. The fingers of the right hand have positive numerical values; those of the left hand, negative values.

Let us suppose that the first player on Team A raises three fingers of his right hand. This means the number $+3$. Now if he crosses his arms and raises four fingers of his left hand, he signals multiplication by -4; this means $3(-4) = -12$. If he next places his hands sidewise across the front of his neck and raises five fingers of his right hand, he is subtracting $+5$ from -12. $(-12) - (+5) = -17$. If he then holds his hands up and raises three fingers of his left hand, he is adding -3 to -17 to get -20. Finally, if he lays his palms crosswise on each other and holds up two fingers of his left hand, he is dividing -20 by -2. The final answer is 10.

Within a short time limit, the first player on Team B must call out the answer to the problem signaled by his opponent. If the answer is right, Team B scores 9 points, that is, 1 point for each of the nine signals used to present the problem. If the solution offered is wrong, or if no answer is forthcoming within the time limit, the first player of Team A must state the solution and show how he arrived at it. If he is correct, he scores 18 points, that is, 2 points for each hand signal.

Naturally, the signaler will try to move as rapidly as possible and to include as many different confusing instructions as he can. (He too has a time limit.)

The players on the opposing teams take turns signaling and solving problems until all players on each team have had a chance to do both. The team with the higher score wins.

Stand Up and Be Counted

Multiplication, Decimals

Twenty-five or more players   Index cards

Paper and pencil   Blackboard and chalk

The players are divided into teams. In a classroom each row can be a team.

Depending on the number of players, everyone receives one or more cards to prepare for the game by writing on each card a number from 0 through 9. Five cards should be prepared for each number, making fifty cards in all.

After the cards have been collected and checked, they are shuffled and dealt out, two to a player. No player should have two cards of the same value. If this happens, players on the same team should interchange cards.

The leader—parent or teacher—now writes an example in multiplication on the blackboard. Say the problem is 345×6.

All the players who have any of these four digits on their cards stand up and show their cards. The leader checks them and credits each team with 1 point for every number shown by a member.

Now the first player of the first team begins the multiplication, saying, "Five times six equals thirty." (If he makes a mistake, his team is penalized 1 point.) All players with the digit 3 or 0 now stand up (note those with 3 are standing up for the second time), and again each team gets 1 point per standee. One of them goes to the blackboard and chalks up the operation so far:

$$
\begin{array}{r}
345 \\
\times\ 6 \\
\hline
30
\end{array}
$$

Next, the first player of the second team continues the multiplication, saying, "Forty times six equals two hundred forty." (He multiplies by 40 because the digit 4, in the tens' position in 345, signifies 4 tens.) All players with 2, 4, or 0 on their cards stand up and score for their team or teams. One of them writes the number 240 underneath the 30 at the blackboard:

$$
\begin{array}{r}
345 \\
\times\ 6 \\
\hline
30 \\
240
\end{array}
$$

Now the first player of the third team says or should say, "Three hundred times six equals eighteen hundred." (He multiplies by 300 because the digit 3 stands in the hundreds' place in the number 345 and denotes 3 hundreds.) Now all the players with the digits 1, 8, or 0 stand up. Since 0 is used twice in the partial product, a bonus of 5 points goes to the team for each of its players who has this digit. Again one of the standees goes to the blackboard and adds the partial product, 1800, to the column:

$$345$$
$$\times\ 6$$
$$30$$
$$240$$
$$1800$$

The first player on the fourth team goes to the blackboard, draws a line under the 1800, adds the four columns, and enters the total, 2070, which is the product. All players with the digits 2, 0, or 7 stand up and score as in the third step, with a bonus of 5 points for the zero, since it is used twice in the answer.

Finally, the first player of the fifth team checks the answer. He divides 2070 by 6 and derives the quotient 345.

For the second round a new example is chalked on the blackboard. This time the second player of the fifth team is followed by the second player of the fourth team, and so on, finishing with the second player of the first team.

As indicated, one point is scored if a player gets a correct partial product $(5 \times 6 = 30, \quad 40 \times 6 = 240, \quad 300 \times 6 = 1800)$. 1 point is deducted per wrong answer, and 5 points are won if the same digit is used twice in an answer.

In later rounds the problems can be made more difficult.

Divide and Conquer

Division, Decimals

Twenty-five or more players   Index cards

Paper and pencil   Blackboard and chalk

In this variation of Stand Up and Be Counted, the same procedure is followed, except that the examples all involve division instead of multiplication.

The long divisions may be of integers—for example, $6794 \div 72$—or may involve decimals—say, $903.75 \div 62.8$.

Each partial quotient is written at the blackboard, so that all the steps in the solution of the problem can be seen by everyone, and the players having the digits used in each step stand up and earn points for their teams.

Down-Count

Subtraction, Fractions, Decimals

Six or more players   Timer

The first player begins with any number and counts down by subtracting the same amount repeatedly. For example, starting from 53, he may count: 53, 50, 47. He stops after three numbers.

Now the next player, within a given time, must continue the countdown from where the preceding player left off, carrying it for three more numbers—in this case, 44, 41, 38. The other players continue, even going on to negative numbers if necessary, until one makes a mistake. A player who gives a wrong response is penalized 1 point, and a player with 3 points against him is eliminated from the round. The winner is the last survivor.

After a few rounds in which the countdown proceeds by whole numbers, the players can try counting down by fractional amounts like 1/2 or 3 3/4. For example, beginning with 22 1/2, they might proceed to 22, 21 1/2, 21, etc. Or, starting with 20, they can proceed to 16 1/4, 12 1/2, 8 3/4, etc.

Finally, the countdown may involve the subtraction of decimals—for example, 0.4. Thus, starting with 27.3, the first player would count 27.3, 26.9, 26.5.

The game can be made increasingly difficult in each round.

Who Am I?

Addition, Subtraction, Multiplication, Division, Fraction and Decimal Operations, Square Roots and Powers

Six or more players   Paper and pencil

One of the players is chosen to be "it." He makes up a problem and presents it to the other players to solve. The one who solves it first, and is declared correct in his work by the others, becomes "it."

For example, the player who is "it" may say: "If you double me, then subtract seven and add three, I will be eight. Who am I?"

Starting with x as the unknown quantity, the players should double it, to produce $2x$, then subtract 7, making $2x - 7$, and finally add 3, ending with the equation $2x - 7 + 3 = 8$. The solution is then easy: $2x - 4 = 8$. $2x = 12$. $x = 6$.

Problems can involve fractions, decimals, squares, square roots, and all the operations of arithmetic. The game can be escalated by increasing the number of operations in the question.

The number of points awarded for solving a problem can be decided by the players according to its difficulty and the number of steps involved in finding the answer. The winner is the player with the highest score after every player has solved at least one problem.

Multicard

Addition, Subtraction, Multiplication, Division

Six or more players, divided into teams

Paper and pencil   Timer

Deck of playing cards

Shuffle the cards and place the pack face down on the table.

The first player of the first team draws three cards, turns them face up, arranges them in any order he pleases, and writes down two different operational signs, one between the first and the second card, the other between the second and the third card, trying to find the arrangement and operations that result in the highest possible multiple of 3.

In this game the jack counts 11; the queen, 12; and the king, 13.

If the player draws 5, 8, and king, he may arrange them as 13 (king), 5, 8. If he writes them as $(13 \times 5) - 8$, this equals $65 - 8 = 57$, or 19×3. So he scores 19 points. But he could score more if he wrote a similar expression of a different arrangement: $(13 \times 8) - 5 = 104 - 5 = 99 = 33 \times 3$, giving a score of 33. A time limit is imposed on the players to sharpen their perception of the possibilities in each combination.

The game continues, with players from opposing teams taking turns, until one team has accumulated 100 points, and wins.

Later rounds can be played with multiples of 4, 5, 6, etc.

Table Building

Addition, Multiplication

Six or more players, divided into teams

Prepared cards   Blackboard and chalk

The only tables built in this game are multiplication tables.

First rule off a large ten-by-ten-box square on the blackboard, as illustrated. Then prepare a hundred cards—ten with the number 1, ten with 2, and so on up to ten with 10. Shuffle the cards and turn the pack face down on the table.

The first player of Team A draws five cards. He may write the numbers he finds on the cards in various vertical columns on the blackboard with the object of building sums that are a multiple of 3, like 6, 9, 12, and 15.

Suppose he draws 5, 7, 3, 2, and 10. He tries two columns:

$$
\begin{array}{cc}
5 & 10 \\
7 & 2 \\
3 &
\end{array}
$$

The first column adds up to 15; the second, to 12 (both multiples of 3). For the two columns he scores 2 points.

In addition to the numbers left on the blackboard, the first player of Team B can work with the five cards he draws, say, 9, 6, 1, 8, and 3. He too can place these numbers in two columns. He adds the 3 to the column already on the blackboard with 10 and 2:

$$
\begin{array}{c}
10 \\
2 \\
3
\end{array}
$$

And he arranges the other four cards in a new column:

$$
\begin{array}{c}
9 \\
6 \\
1 \\
8
\end{array}
$$

One of his columns now adds up to 15, and the other to 24.

163

Both numbers are multiples of 3. He too scores 2 points for his two columns.

The game proceeds until all the cards have been drawn. High score wins. Cards drawn that are not usable by the drawer are to be placed face up on the table, to be taken by any player who can use them.

The next rounds of the game can be played with multiples of 4, 5, 6, etc. The game can also be made more challenging by awarding 1 point for a column of two numbers adding up to the required multiple, 2 points for a column of three, 3 points for a column of four, etc.

5	10	9							
7	2	6							
3	3	1							
		8							

Rank and File

Addition, Multiplication

Six or more players, divided into teams

Blackboard and chalk   Deck of playing cards

In this variation of Table Building, rule off a twenty-five-box square on the blackboard, as illustrated.

On the right side of the ranks or horizontal rows place five numbers, not necessarily in sequence, in ascending order from the top down to represent different multiplication tables. At the top of the files or vertical columns place the same numbers in ascending order from left to right. In the square illustrated here we have used 6, 7, 8, 9, and 12. (In other rounds other numbers can be substituted.)

Place in the center box any number from 1 through 9. In the illustration it's a 9. (This number is changed in every round.)

The ace can count either 1 or 15; the jack, 11; the queen, 12; and the king, 13.

In the first round the first team will add its numbers down the columns, and the second team along the rows. After each round the teams reverse this arrangement.

After the cards have been shuffled, the first player of the first team draws two—let us say, a 9 and a king. He enters the numbers anywhere on the board—in the same column or in different columns. In the game illustrated, he places the 9 in the seven column. In placing it also in the six row, he gives his opponents both a problem and an opportunity, as we shall see. He places the 13 in the nine column and in the seven row.

Now the first player of the second team also draws two cards. If he too draws a king and a 9, he can place both numbers, as illustrated, in the seven row. He then can add $13 + 13 + 9 = 35 = 5 \times 7$. His score is not 35, but 5, which is the multiple of 7, the rank in which he added his figures. He thus took advantage of the opportunity provided by the 13 already placed in the seven rank by his opponent.

The numbers remain on the blackboard as the second player of the first team draws two cards. If he draws, say, a 6 and a queen (12), he can enter both numbers in the eight column as shown. Now he has a sum in that column: $13 + 9 + 12 + 6 = 40 = 5 \times 8$, scoring 5. Note that he took advantage of the 9 in the center box and that he added the adjacent numbers in the eight file needed to make a sum that is a multiple of 8.

If his counterpart on the second team next draws a 10 and a 5, he can enter both numbers on the eight row, as illustrated. His sum is then $5 + 10 + 9 = 24 = 3 \times 8$, and his score is 3.

Now suppose the third player on the first team draws a jack (11) and an 8. He enters the 8 in the eight column; he increases the sum in that column from 40 to $8 + 13 + 9 + 12 + 6 = 48$; and his score is 6. As for the 11, he decides to place it in a box that will not be likely to help his opponents — in the twelve column and also in the twelve row.

166

The third player on the second team may draw a 4 and a queen (12). Entering them both in the eight row, he increases the sum in that row from 24 to $5 + 10 + 9 + 4 + 12 = 40 = 5 \times 8$, with a score of 5. The first team now has 11 points, and the second team has 13.

The game continues until all the boxes are filled. Several rounds should be played until one team has accumulated 100 points.

6	7	8	9	12	
	9	8			6
		13	13	9	7
5	10	⑨	4	12	8
		12			9
		6		11	12

Multibingo

Multiplication

Six or more players   Two dice

Paper and pencil

Each player prepares a twenty-five-box square and inserts in the boxes at random twenty-four numbers selected from

167

the following eighteen, with six numbers repeated: 1, 2, 3, 4, 5, 6, 8, 9, 10, 12, 15, 16, 18, 20, 24, 25, 30, 36. Four such boxes are illustrated.

The eighteen numbers are the possible products of the numbers turned up when two dice are thrown. The players take turns rolling the dice and multiplying (silently) the two numbers turned up. The teacher or leader keeps a record of these numbers. Each player checks off the box in his square that contains the product of the two numbers. Thus, if a player rolls a 5 and a 4, he and all the other players should check off the number 20, wherever it may

8	20	30	2	6
10	1	12	5	18
15	3	✕	24	10
25	30	4	18	16
1	8	12	15	30

2	16	10	4	24
12	8	36	3	9
25	4	✕	6	15
5	15	1	20	30
30	18	9	24	3

12	30	6	36	16
25	2	10	4	24
20	8	✕	18	10
3	15	3	24	1
2	12	5	16	9

6	12	25	2	15
9	3	18	36	25
5	16	✕	8	10
24	1	5	4	2
10	8	30	15	20

be. A combination of 5 and 3 calls for all the players to check off 15.

The game continues until one player has checked off all the numbers in a row of five, a column of five, or a diagonal. (These can all include the "wild" center square.) He is the winner.

If a product that has already been checked off recurs, the dice pass to the next player.

Of course, the player calling "Bingo!" should have his card checked by the leader or the teacher to make sure that his multiplications were correct.

Mathematical Ping-Pong

Addition, Subtraction, Multiplication, Division, Fractions, Decimals, Equations

Six, eight, ten, or larger even numbers of players

Timer

The players form two teams. Player 1 on Team A "serves" an arithmetical "ball" consisting of an expression that involves addition, subtraction, multiplication, division, fractions, decimals, or any combination of them. It is best to start with something simple, like 5×5.

Within an agreed time limit, Player 1 on Team B must step forward and "return" an exact mathematical equivalent of the first expression, but without repeating the expression or stating what number it equals. For example, he may say, "Half of fifty."

This equivalent expression must then be "hit" back, and so on, always within the time limit, without any exact repetition of expressions already used.

To confuse his opponent, a player will try to make the expression as complex as possible — at the same time, of course, giving himself a little intellectual exercise.

The game might continue with more equivalents of 25, escalating all the while:

Player 2, Team A: Open parenthesis, eight times two, close parenthesis, plus nine.

The expression to be visualized by the players on Team B is $(8 \times 2) + 9$.

Player 2, Team B: Wow! Take this: Open parenthesis, eighty minus five, close parenthesis, divided by three.

Player 3, Team A: OK. Twenty-six and one-half minus one and one-half.

Player 3, Team B: Thirty minus six point five plus three point two minus one point seven.

1 point goes to the team whose opponent cannot return the ball within the time limit, and 2 points are earned for correctly pointing out that a player on the opposing team

has repeated an expression already used or has made an error.

In either case, a new combination, with a different value, is then served, and the game proceeds until players on one team have accumulated a total of 10 points.

Mathematical Seesaw

Addition, Subtraction, Multiplication, Division, Fractions, Decimals, Squares, Square Roots, Equations

Six, eight, or larger even numbers of players

Paper and pencil   Timer

In this variation of Mathematical Ping-Pong, the level of difficulty is escalated. It is guaranteed to give the players plenty of mental exercise.

Divide the group into two teams, A and B.

A member of Team A tips the seesaw by announcing an arithmetical, algebraic, or geometrical expression equivalent to a certain "weight" in pounds—for example, "Open parenthesis, ten times two cubed divided by four, close parenthesis, plus the square root of twenty-five."

Now the first member of Team B must, within a given time limit, "balance" the seesaw by putting an equivalent weight upon it, but expressed in other terms — say, "The length of the hypotenuse of a right triangle having fifteen for one side and twenty for the other."

The teams then take turns, player by player, trying to keep the seesaw always unbalanced (Team A) or in balance (Team B) with different combinations resulting in the same number of pounds. In this case, the number is 25. The game can proceed with expressions involving decimals, fractions, the volumes of solids like triangular pyramids, the lengths of tangents to a circle, or the application of any known formulas, depending on the ability and maturity of the players.

A team scores 1 point if a member of the opposing team cannot, within the time limit, produce an expression of equivalent "weight"; and 2 points are scored for one's team for correctly pointing out that a member of the opposing team has either repeated an expression already used or made an error.

In either case, a new combination is chosen, with a different "weight," and the game continues until one team has accumulated 10 points.

As can be seen, the players really have to weigh their words!

Measure for Measure

Equivalences of Linear, Liquid, Time, Weight, Volume, and Monetary Measurement, Geometric Figures

Six or more players   Timer

The first player, beginning with a small whole number — say, 2 — makes a statement of measurement or mathematical relationship involving that number, for example, "There are two pints in a quart."

Now the next player must, within a given time, match this statement with another involving the identical number, for example, "There are two quarters in a half dollar."

The game might continue:

Player 3: There are two half dollars in a dollar.

Player 4: There are two nickels in a dime.

Player 5: There are two sides to an angle.

Player 6: There are two equal sides in an isosceles triangle.

Player 1: There are two equal angles in an isosceles triangle.

Player 2: There are two diagonals in a square.

Player 3: There are two weeks in a fortnight.

Player 4: There are two parts to a bisected line.

Player 5: There are two congruent parts to a bisymmetrical figure.

Player 6: There are two terms in a binomial.

Player 1: There are two points that determine the length of a straight line.

Player 2: There are two points on the circumference of a circle that determine an arc.

Player 3: There are two coordinates that determine the location of a point in a plane.

Player 4: There are two digits in every integer from 10 through 99.

Player 5: There are two prime numbers between 13 and 23 [17 and 19].

Player 6: There are two parts to a fraction [numerator and denominator].

Player 1: There are two roots in a quadratic equation.

Player 2: I'm stuck. You score one point. I'll begin with three. There are three sides to a triangle.

The game goes on this way, with the players raising the number by one each time someone repeats a response already given or fails to supply a new one within the agreed time, or errs in doing so. Players can mention the three

174

equal sides in an equilateral triangle, three angles in a triangle, three miles in a league, three feet in a yard, then the four quarters in a dollar, four quarts in a gallon, four sides in a quadrilateral (whether square, rectangle, or trapezoid), four pecks in a bushel, four gills in a pint, four fluidounces in a gill, etc.

The player who first scores 5 points wins the game.

Step Forward!

Addition, Subtraction, Multiplication, Division

Eighteen or more players, divided into teams

Paper and pencil   Two dice

Prepared cards and tags

First each team prepares number tags from 0 through 9 to be worn by the individual players. If there are more than nine players on each team, some of the numbers may be repeated.

Each team should also have five sets of four cards, each marked with the symbol of an arithmetical operation: $-$, $+$, $\div$, and $\times$, representing subtraction, addition, division, and multiplication. Additional cards can be marked with open parenthesis (and close parenthesis). Two cards should be marked with an equal sign: $=$.

Player 1 of Team A now rolls the dice. Say he throws 6 and 3. These numbers can be read as 63 or as 36.

Using "MADS" (multiplication, addition, division, and subtraction), the first team must try to use as many players as possible to produce the number agreed on. For example, suppose the number is 36. Then $5(9 - 2) + (8 \div 4) - 1 = (5 \times 7) + 2 - 1 = 37 - 1 = 36$. Here the team made use of the symbols $\times$, $-$, $+$, $\div$, and $-$. (No more than one operational symbol may be repeated. Here the symbol $-$ was used twice.) The parenthesis signs were used as needed.

The team scores 1 point for each number used to produce the result required. Here six numbers (5, 9, 2, 8, 4, and 1) were used. So Team A scores 6 points.

Now suppose Player 1 of Team B rolls a 4 and a 3. This could be 43 or 34. The team offers $(8 \times 5) + (9 - 7) + 1 = 40 + 2 + 1 = 43$. Here five numbers (8, 5, 9, 7, and 1) are used, and the team scores 5 points.

The game continues until one team has accumulated 25 points.

Card Board

Addition, Subtraction, Multiplication, Division

Six or more players   Prepared cards

Paper and pencil

The cards prepared for this game should be small enough to fit in the boxes of a board based on the one illustrated below.

Prepare ten cards with each of the signs $+$, $-$, $\times$, and $\div$, or forty cards in all. Then prepare thirty cards with the sign $=$. For each of the numbers 0, 1, 2, and 3, prepare seventeen cards; for each of the numbers 4 and 5, eleven cards; for each of the numbers 6 and 7, eight cards; and for each of the numbers 8 and 9, seven cards, making 120 cards in all with numbers on them.

Place the numbered cards, after they have been shuffled, face down on the left side of the board. Set the cards with the operations marked on them, after they too have been shuffled, face down on the far side of the board. Lay the equality cards, face up, on the right side of the board.

Each player draws six number cards from the top of the pile, two operational cards, and one equality card. The object of the game is to place cards on the board to form true mathematical statements, and to arrange them so as to obtain the maximum number of points.

Note that the board has been arranged with certain boxes colored (or marked) black, green, orange, and red. If a card with a number forming part of a true mathematical statement is placed on a red box, the score of the whole statement is doubled. A number card in a green box scores double the value of the number on the card. If a number card is placed in an orange box, it scores triple. A number card in a black box squares the value of the number in the calculation of the score. In the other boxes the score for a number card is the value of the number. An equality sign scores 0; a plus card scores 2 points; a minus card scores

3 points; a multiplication card scores 4 points; and a division card scores 5 points.

Let us examine a sample game.

The first player lays out as many cards as will form a true statement (if he can), arranging them either horizontally — e.g., $3 + 5 = 4 \times 2$ — or vertically:

$$
\begin{array}{c}
3 \\
+ \\
5 \\
= \\
4 \\
\times \\
2
\end{array}
$$

Naturally, he tries to place the statement so as to score as many points as possible. Let us say he places the card with the number 4 in a red box. He scores 2 points for the +, 4 points for the ×, and 14 points for the numbers on the cards. The total of 20 points is multiplied by 2 because one of the number cards is in a red box. So his total score is 40 points.

The next player may make a true mathematical statement by appending his cards to any part of the statement already on the board, whether horizontally or vertically. For example, he may add on a 4, an ×, an =, a 9, a +, and a 3, as follows:

$$
\begin{array}{l}
4 \\
\times \\
3 + 5 = 4 \times 2 \\
= \\
9 \\
+ \\
3
\end{array}
$$

178

Each player should plan his moves carefully to take advantage of the extra points he can score by placing his cards in the colored boxes. If a player challenges the correctness of an equation laid down on the board and can point out an error in it, it must be removed, and the player who put it down loses his turn. After a player has laid down an equation, he replenishes his stock of cards by drawing new ones from the three groups on the table so that he again has six number cards, two operational cards, and one equality card.

		B		R				O					
			G			R	G						B
B	O		G	G			R			O			
	G	O						O	R				
G			O		G		O					G	
	O					G							
R			B				G						
	B				R				B				
			G				B				R		
				G					O				
G			O		G		O				G		
	R		O					O		G			
	O			R		G	G			O	B		
B			G	R				G					
		O				R			B				

The game ends when all the cards have been drawn from the piles on the table or when a stalemate occurs, that is, a situation in which no player can make a move. The highest accumulated score wins.

When it is difficult to find a way of attaching cards to those already on the board, a player may be able to use a combination of ÷ and 1 or × and 1. For example, suppose that the equation $8 + 6 = 7 \times 2$ is on the board. A player can add ×1 to this, making it $8 + 6 = 7 \times 2 \times 1$. Although the value of the equation is not changed, the player who adds ×1 (or ÷1) to it can score for the entire line: the sum of all the numbers used $(8 + 6 + 7 + 2 + 1 = 24)$ and of the operational values of + (2), and × (4 + 4) — a grand total of at least 34, not counting additional points that might be scored if the number 1 is placed in a colored box.

Count Up!

Counting, Addition, Subtraction, Multiplication, Division, Fractions, Decimals, Exponents, Finding Square Roots, Fibonacci Numbers

Six or more players, in teams   Timer

In the simplest form of the game, the first player of the first team begins the count with the number 1. Within a given time limit, he makes as many true statements as he can, involving mathematical operations, that result in 1. For instance, he may mention 1/5 of 5, $6 - 5$, 3^0, 1^2, etc. For each correct response given within the time limit, he earns 1 point for his team. On the other hand, his team is penalized 1 point for each wrong or repeated response. (This includes *kind* of response, e.g., 1^2 and 1^3, or 3^0 and 101^0.)

The first player on the second team then counts up to 2 and starts making as many true statements as he can about 2, such as 1/4 of 8, $4 - 2$, $4/2$, $6.3 - 4.3$, $\sqrt{4}$, and $1 + 1$. For the number 3, statements might include $5 - 2$, $6/2$, 3×1, $2 + 1$, and $2^2 - 1$; for the number 4, 2^2, $2 + 2$, and $(3 \times 2) - 2$; for the number 5, $8 - 3$, $4 + (3 - 2)$, and $2^2 + 1$.

Each member of each team gets a turn. The team with the greater number of points scored wins.

There are a number of ways in which the difficulty of the game can be increased. After all players have had a turn, a new round can be begun with the next higher number instead of returning to 1. For example, if there are sixteen players, the second round can begin with the number 17. Or the counting up can be by powers of 2: 1, 2, 4, 8, 16, . . . ; or powers of 3: 1, 3, 9, 27, ... Finally, players may count up along the basic Fibonacci series, 0, 1, 1, 2, 3, 5, 8, 13, . . . , in which each new term is produced by the addition of the two last terms. Start counting up with the second 1.

181

Triangle Stretch

Plotting Points on a Grid, Calculating the Area of a Rectangle and a Triangle, Addition, Subtraction

Six or more players, divided into teams

Pencil   Red and white dice

Graph paper

Each team prepares a grid of six vertical and six horizontal lines, numbered from left to right at the base and from the base to the top at the left, as illustrated.

The first player of the first team rolls the dice three times. The numbers on the red die denote vertical lines; the numbers on the white die, horizontal lines. Thus, each roll, such as red 1, white 2, can be used to plot a point at which the corresponding lines intersect — in this case, the crossing of the first vertical line with the second horizontal line. Three such points should be plotted and marked on the grid. The second point should not be on the same vertical or horizontal line as the first. So if a player rolls a point that he has already marked or one which forms a straight line with the first two, he disregards it and rolls again. A player continues to roll the dice until he can plot a point that is not on that line. He then connects the three points to form a triangle.

Suppose, for example, that the first three rolls are red 1, white 2; red 3, white 5; and red 5, white 1. The resulting triangle, formed after the three points plotted on the grid are connected, is illustrated.

182

The player must now compute the area of the triangle. To do so, he "stretches" it to form a rectangle by drawing vertical and horizontal lines through the vertexes of the triangle, as illustrated. Labeling the smaller triangles formed within the rectangle as I, II, and III, he calculates the area of the large triangle (marked T) by means of the formula: area of T = area of Rectangle − (area of I + area of II + area of III).

The area of a rectangle is equal to its length multiplied by its height. The area of a triangle is equal to one-half the product of the length of one side and the height of a perpendicular dropped to that side from the opposite vertex.

Substituting these formulas in the equation above, the area of the rectangle is $4 \times 4 = 16$. The area of triangle I is $\frac{1}{2}(3 \times 2) = 3$; of triangle II, $\frac{1}{2}(4 \times 2) = 4$; of triangle III, $\frac{1}{2}(4 \times 1) = 2$. Then the area of triangle T is $16 − (3 + 4 + 2) = 7$.

This is the score credited to the first team.

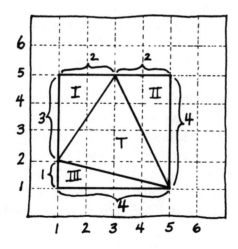

The teams take turns until one team has accumulated a previously agreed-on number of points — say, 25.

Of course, a new grid must be prepared for each turn.

Metric Baseball

Metric Conversions

Six, eight, or larger even number of players

Paper and pencil	Disks
Prepared cards	Timer

First draw a baseball diamond.

Then divide the players into two teams. Each team receives twenty to thirty-two cards, depending on the number of players, in four equally numerous categories. If twenty cards are used, five should be marked "single"; five, "double"; five, "triple"; and five, "home run."

The players on each team now make up problems in metric conversion and decide among themselves on the value to be assigned to the solution of each problem, depending on its difficulty. For example, solving the easy problem of converting 5000 grams to kilograms might be valued as a single. Converting 22 pounds to kilograms might be worth a double. Changing 40 kilometers to miles could be equivalent to a triple. And finding the number of grams in 1 pound, 8 ounces, could be valued as a home run.

Each problem is written on a card corresponding to its agreed value, e.g.: 5000 grams = x kilograms. There should be an equal number of cards of all four levels of difficulty.

The players next choose to see which team is at bat first. When Team A is at bat, the cards of Team B are shuffled thoroughly and placed face down on the table. The first player of Team A then chooses a card from this pile. Within a given time limit he must solve the problem on the card. For example, if it is the home-run problem of converting 1 pound, 8 ounces, into grams, he adds 453.6 grams (the equivalent of 1 pound) and 226.8 grams (the equivalent of half a pound or 8 ounces; see Table of Metric Equivalents in the Appendix). The answer is 680.4 grams, rounded to 680 grams. If the player can arrive at a correct solution within the time limit, Team A is credited with the value of the problem—a home run. Otherwise his team is out, and Team B goes to bat. (In this game there is only one out per inning.)

Each team draws cards from the pile of the opposing team. As hits are made, disks are placed on the diamond to show the positions of players at first, second, or third base. Players advance from base to base, as in baseball, as hits are made by members of their own team; a player on second base advances only to third base on a single by a teammate.

The game should be played for five innings or as long as there are cards in the piles of both teams. The team with the higher score wins.

Equation Circulation

Solving Algebraic Equations

Six or more players

Paper and pencil

Prepared cards

In this game the equations to be solved circulate among the players according to their mathematical abilities.

Write the numbers from 1 through 10 on ten cards and shuffle.

The first player draws two cards and holds up one of them for the rest to see. Say he draws a 5 and an 8 and holds up the 8, concealing the 5. He then makes up and states a simple algebraic equation with 8 on the right-hand side and the unknown value on the left — for example, $2x - 2 = 8$. (The solution is $x = 5$, the concealed number.) He asks, "What is my other number?"

The first player to solve the equation raises his hand and offers his solution, showing how he worked it out. If he is correct, he scores 1 point and it is his turn to draw two cards, etc.

The player who first accumulates 10 points wins.

Count on Me

Multiplication and Division of Whole Numbers, Fractions, and Decimals

Six or more players, divided into teams

Timer

In this game each player "counts on" the numbers called out by the previous player by multiplying or dividing them, as the case may be, by the same amount.

Player 1 of Team A sets the pattern by calling out three numbers — say, 4, 8, and 16. Within a given time limit, Player 1 on Team B, after noting that each number called is twice the preceding one, continues multiplying by 2 to get, and call, 32, 64, and 128. If Team A's Player 1 began with 500, 100, and 20, Team B's Player 1 would continue to divide by 5, calling 4, 4/5, and 4/25.

The players of each team take turns continuing the series until a player makes a mistake. His team is penalized 1 point, and a new series begins. A player with 3 points against him is out of the game. The winning team is the one with the last surviving player.

Somewhat more difficult is continued multiplication by a fraction. In the series 75, 30, and 12, the multiplier is 2/5. Or you can say each term is divided by 2 1/2. Once the players develop skill, they should have no difficulty in dividing numbers by a fraction, because this just involves inverting the fraction and multiplying. Thus, the series 90,

120, 160 is produced by dividing each number by 3/4, which is equivalent to multiplying by 4/3.

Also, decimals can be used to multiply or divide in counting up or down.

Figure on Me

Solving Algebraic Equations

Six or more players, divided into teams

Paper and pencil   Timer

Prepared cards

Prepare twice as many cards as there are players, writing a number on each card in sequence: 1, 2, 3, 4, etc.

Shuffle the cards, deal out half of them, and pin one card to the back of each player, so that he can see the number on the back of every other player but not the one on his own.

Each player, in turn, turns his back and tries to determine his number by asking a question about it, using an algebraic expression or equation that can be answered yes or no. A player can ask only one question at a time and must wait for his turn to come again before asking another question.

For example, a player may ask if his number is greater than $x + 5 = 9$. (The answer is yes if his number is 5 or more;

no if his number is 4 or less.) Or he may ask if his number is greater than $2x - 6$, or greater than $\frac{x}{3} = 2$, or less than $5x - 2x = 9$, or less than $x^2 = 64$.

Responses must be given within a given time limit and must be correct, or there is a 1-point team penalty.

The first player to guess his number earns 1 point for his team. The turns keep going around until all players have guessed their numbers. The team with the higher score wins.

In other rounds the numbers pinned on the players' backs can be fractions or decimals, or may be represented as squares or square roots.

Fencing Match

Solving Algebraic Equations

Even number of ten or more players

Blackboard and chalk   Timer

The only fencing that takes place in this game is the fencing in of adjacent digits on the blackboard by the members of the competing teams as they solve algebraic equations.

First, the leader — parent or teacher — writes on the blackboard a row of digits from 0 through 9 in random order — for example, 6, 3, 9, 1, 0, 5, 7, 4, 8, 2. He writes five or six such rows below each other, each row in a different order, and can add more rows, or erase rows, as the game proceeds.

Next, he writes an equation having a solution consisting of digits that are adjacent in a row on the blackboard—let us say, $2x + 1 = 39$.

Player 1 on Team A must, within a given time limit, respond with the correct answer—in this case, 19—or his team loses 2 points. He then goes to the blackboard and fences in the adjacent digits 1 and 9, as shown below. He scores 1 point for his team for each digit fenced in—in this case, 2 points.

A new equation is then chalked on the blackboard—say, $\frac{2x}{3} - 7 = 5$. Player 1 on Team B calls out 18 and goes to the blackboard to fence in the 1 and the 8:

$$7 \boxed{1\ 9} 3\ 5\ 2\ 6\ 0\ 8\ 4$$
$$3\ 0\ 9\ 4\ 6\ \boxed{1\ 8}\ 2\ 7\ 5$$

The game continues in this fashion, and more digits are fenced in. At any point, the numbers on the blackboard may be erased, and new arrangements may be substituted.

The algebraic equations may involve addition, subtraction, multiplication, division, squares, square roots, cubes, and fractions, graded in difficulty according to the abilities of the players.

The team that first accumulates 10 points wins.

190

Here are a few more equations and digit combinations, with the solutions fenced in.

$$x^2 - 11 = 25$$
$$x^2 + 3x - 10 = 450$$
$$\frac{x}{10} + \frac{x}{5} = 96$$
$$\sqrt[3]{x} - 2 = 3$$

4 3 7 9 8 2 5 1 $\boxed{6}$ 0

9 8 $\boxed{2\ 0}$ 5 3 6 7 1 4

7 9 5 8 $\boxed{3\ 2\ 0}$ 4 6 1

4 3 9 $\boxed{1\ 2\ 5}$ 8 0 7 6

Algebraic Locus-Pocus

Plotting Points on a Grid, Solving Algebraic Equations, Solving Simultaneous Equations

Six or more players, divided into teams

Paper and pencil   Timer

Prepared cards

Prepare thirty-six cards with the equations shown here, but without the answers:

1. $x + 3 = 4$; $y + 8 = 9$ ($x = 1$; $y = 1$).

2. $2x - 3 = -1$; $\frac{y}{2} + 3 = 4$ ($x = 1$; $y = 2$).

3. $2x + 3x - x = 4$; $\frac{y}{3} - 9 = -8$ ($x = 1$; $y = 3$).

4. $x + 2 = 3$; $4y - 5 = 11$ ($x = 1$; $y = 4$).

5. $5 - x = 4$; $2y - 6 = 4$ ($x = 1$; $y = 5$).

191

6. $3x - 2x + 4x = x + 4; \frac{y}{2} + 6 = 9$ $(x = 1; y = 6)$.

7. $5x = 10; y - 2 = -1$ $(x = 2; y = 1)$.

8. $3x - x = 4; \frac{y}{2} = 1$ $(x = 2; y = 2)$.

9. $2x + 5x = 3x + 8; 3y - 1 = 8$ $(x = 2; y = 3)$.

10. $\frac{x}{2} + \frac{3x}{2} = 4; 2y + y - 5 = 7$ $(x = 2; y = 4)$.

11. $-5 - 2x = -9; 5y - 4 = 21$ $(x = 2; y = 5)$.

12. $4x - 3 = 5; 2y + 7y - 4 = 50$ $(x = 2; y = 6)$.

13. $7x - 5 = 16; 2y + 8y = 3y + 7$ $(x = 3; y = 1)$.

14. $5x - 2 = 13; 2y + 4 = 3y + 2$ $(x = 3; y = 2)$.

15. $3x - 7 = 2; 2y + 11 = y + 14$ $(x = 3; y = 3)$.

16. $2x + 7x - 4 = 23; \frac{y}{2} + 3 = 5$ $(x = 3; y = 4)$.

17. $7 - 3x = -2; 5y + 10y = 75$ $(x = 3; y = 5)$.

18. $3x - 7 = 2; 5y - 3y = 12$ $(x = 3; y = 6)$.

19. $y + x = 5; y - x = -3$ $(x = 4; y = 1)$.

20. $x = 2y; x + 2y = 8$ $(x = 4; y = 2)$.

21. $2x - y = 5; x + y = 7$ $(x = 4; y = 3)$.

22. $2x - y = 4; x = y$ $(x = 4; y = 4)$.

23. $x - y = -1; 2x + 3y = 23$ $(x = 4; y = 5)$.

24. $2x = y + 2; x + y = 10$ $(x = 4; y = 6)$.

25. $2x + 3 = x + 8; 5y - 3y = 2$ $(x = 5; y = 1)$.

26. $3x - 7 = 8; \frac{y}{2} + 8 = 9$ $(x = 5; y = 2)$.

27. $7x + 2 = 3x + 22; 3y - 7 = 2$ $(x = 5; y = 3)$.

28. $3x + 7 = 2x + 12; 32y = 128$ $(x = 5; y = 4)$.

29. $\frac{x}{2} + \frac{7x}{2} = 20; 10 - 2y = 4y - 20$ $(x = 5; y = 5)$.

30. $3x - 7 = x + 3; 6y - 10 = 26$ $(x = 5; y = 6)$.

31. $5x - 11 = 19; y - 8 = 2y - 9$ $(x = 6; y = 1)$.

32. $\frac{x}{2} - 1 = 2; 2y + 4 = 8$ $(x = 6; y = 2)$.

33. $5x - 2 = 28; 7y - 2y = 15$ $(x = 6; y = 3)$.

192

34. $\frac{x}{2} + \frac{x}{3} = 5$; $12y - 4 = 44$ ($x = 6$; $y = 4$).

35. $3x + 7 + 5x = 55$; $\frac{y}{5} + 7 = 8$ ($x = 6$; $y = 5$).

36. $2x - 3 = 3x - 9$; $2y + 7 = y + 13$ ($x = 6$; $y = 6$).

Next prepare a thirty-six-box square, as illustrated. Shuffle the cards and lay them face down on the table.

Player 1 on Team A draws one card and, within a given time limit, solves the two equations and plots the points (here shown as boxes on the grid) corresponding to the answers, x along the horizontal axis, and y along the vertical axis.

Player 1 on Team B does the same, and all the players of each team keep taking turns until one team gets a pattern of X's or O's, as the case may be, arranged in three contiguous boxes, whether horizontally, vertically, diagonally, in the shape of an L or of a U, etc. (See the three X's circled below.) If a player fails to solve the equations in the allotted time or gives a wrong answer, a player on the opposing team can challenge him, correctly solve the equations, and plot the point for his own team.

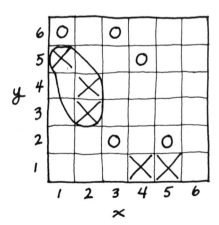

Transfigurations

Addition, Subtraction, Multiplication, Division, Fractions, Squares, Square Roots, Factorials

Six or more players, divided into teams

Timer

A transfiguration is a transformation, as of something familiar into something unfamiliar, or vice versa. With a little imagination, figures can be transfigured.

Let us, for example, begin with the figure 2. Player 1 on Team A tries to find three transfigured 2's that equal 0, within a given time limit.

He may arrive at this: $(\sqrt{2})(\sqrt{2}) - 2 = 0$.

Now Player 1 on Team B must make three transfigured 2's that equal 1.

He might solve the problem thus: $\dfrac{(\sqrt{2})(\sqrt{2})}{2} = 1$.

Player 2 on Team A (if there are two teams) has to use three transfigured 2's to equal 2. Say he cannot solve the problem in time; Player 2 on Team B is given a chance and might offer $2 - 2 + 2 = 2$. (There are three 2's on the left side of the equation.) The players can continue with $2 + \dfrac{2}{2} = 3$, and $2 + (\sqrt{2})(\sqrt{2}) = 4$. When no player on either team can carry on the series, a new series begins, with

194

three transfigured 3's: $(\sqrt{3})(\sqrt{3}) - 3 = 0$; $\dfrac{(\sqrt{3})(\sqrt{3})}{3} = 1$;

$\dfrac{3+3}{3} = 2$; $3 \times 3 \div 3 = 3$; $3 + \dfrac{3}{3} = 4$; $3! - \dfrac{3}{3} = 5$. (3! is called

"factorial 3" and is equivalent to $3 \times 2 \times 1 = 6$; $4! = 4 \times 3 \times 2 \times 1 = 24$, etc.)

At a higher level, the game can be played with sets of four transfigured figures. Beginning with four transfigured 2's, the players might proceed as follows:

$$22 - 22 = 0$$

$$\frac{22}{22} = 1$$

$$\frac{2}{2} + \frac{2}{2} = 2$$

$$(2 \times 2) - \frac{2}{2} = 3$$

$$2^2 - 2 + 2 = 4, \text{ or } \left(2 \times \frac{2}{2}\right)^2 = 4$$

$$2^2 + \frac{2}{2} = 5$$

$$2^2 + (\sqrt{2})(\sqrt{2}) = 6$$

$$2 \times 2 \times (\sqrt{2})(\sqrt{2}) = 8, \text{ or } \frac{(2^2)^2}{2} = 8$$

$$\frac{22}{2} - 2 = 9$$

$$\frac{22}{(\sqrt{2})(\sqrt{2})} = 11$$

$$\frac{22}{2} + 2 = 13$$

$$2 \times 2 \times 2 \times 2 = 16$$

$$22 - (2 \times 2) = 18$$

Can you supply transfigurations for any missing numbers above?

A team scores 1 point when a member solves a problem that a player on an opposing team has failed to solve within the allotted time.

The team with the higher score after ten rounds is the winner.

As can be seen, the possibilities of transfigurations are many — but there are also impossibilities, as you will see when you play the game.

PART TWO

*Mathematical Wizardry
and Wit*

6

Mathemagic

THE YOUNGSTER who can fascinate his friends with seemingly magical feats of mathematical prowess gains not only prestige among his peers but a deeply satisfying confidence in his own powers. There is great fun in being able to solve instantaneously what look like highly complex mathematical problems and in showing off one's apparently extraordinary skill in the performance of calculations with the speed and accuracy of an electronic computer.

The "mathemagic" in this chapter can turn any child into a veritable Houdini with numbers and a featured attraction at parties and social gatherings.

Age Sage

Announce that you can tell everyone present not only his age but the exact month and day of his birth. That is because you are an "age sage," a person endowed with special mathematical wisdom that enables you, in a few minutes, to perform the calculations needed to get this information about any person.

There is no doubt that your challenge will be taken up. But you need not worry: you will meet it easily — to the astonishment of everyone.

Each person present should be directed to write down and conceal a three- or four-digit number composed of the month and day of his birth. (If he was born on the first of August, the number is 801, not 81.) He is to multiply this figure by 2, add 5, multiply by 50, add his age in years, add 365, subtract 615, and tell you his answer. Instantly you announce the month and day of his birth and tell him how old he is!

Try this with a child of twelve who was born on August 20. Following your directions he writes:

Month and day of birth:	820
Multiply by 2:	1640
Add 5:	1645
Multiply by 50:	82,250
Add age in years — in this case, 12:	82,262
Add 365:	82,627
Subtract 615:	82,012

The number 82,012 consists of 8 (August), 20 (day of the month), and 12 (age of child).

Mathematical Telepathist

A mathematical telepathist is a fancy and impressive name for a person who, without asking any questions, can give the answer to another person's mental calculations that start with a number unknown to the telepathist.

You can be a mathematical telepathist. Begin by asking a friend to think of a number. Tell him to double it. Then have him add, say, 12, and divide the result by 4. Finally, he is to subtract half the original number. You tell him what his answer is, and he wonders how you could possibly have found it. Perhaps it was just luck. So you give him another round, starting with a different number. This time, however, you tell him, after doubling the number, to add, say, 20 before dividing by 4 and subtracting half the original number. Again you tell him his precise answer. You are indeed a mathematical mind reader!

What your friend does not know, of course, is that the whole secret of your prowess lies in the number you told him to add after doubling his original number. All you have to do to determine the answer to his calculations is to divide that added number by 4. So, if you told him to add 12, his answer will be 3; and if you told him to add 20, his answer will be 5. Just remember always to tell him to add a number that you know is divisible by 4.

The algebra is: he starts with x, doubles it, and adds $4k$. Dividing by 4 leaves $\frac{x}{2} + k$, and subtracting $\frac{x}{2}$ leaves just k.

Let's see how it works when you add, say, 16.

Suppose your friend starts with 8.

$8 \times 2 = 16$. $16 + 16$ (the added number) $= 32$. $32 \div 4 = 8$.

$8 - \dfrac{8}{2} = 8 - 4 = 4$. $4 = 16 \div 4$.

Psychic Sleuth

You let it be known among your friends that you have the uncanny ability to detect, by means of psychic mathematical powers, the exact number of people in anyone's family.

If, at a gathering, you are asked to demonstrate your ability as a psychic sleuth, you hand out paper and pencil to everyone. Each person present performs a few simple arithmetic calculations and lets you know the result. You then inform each one how many brothers and sisters he has and the number of his living parents, to the astonishment of all.

Let b = the number of brothers; s = the number of sisters; and p = the number of living parents.

The instructions you give are as follows:

First, multiply the number of brothers by 2. (If you keep your own record, it would then show $2b$. Of course, the letters in your record will represent various numbers in the calculations of the different persons present.)

Next, add 3. This gives you $2b + 3$.

202

Now, multiply by 5. The result is $5(2b + 3) = 10b + 15$.

Then, add the number of sisters. This is $10b + 15 + s$.

Multiply by 10. This is $10(10b + 15 + s) = 100b + 150 + 10s$.

Add the number of living parents. This gives $100b + 150 + 10s + p$.

Finally, subtract 150. The result is $100b + 10s + p$. In short, the three-digit number constituting a person's answer is to be interpreted as showing the number of brothers in the hundreds' place, the number of sisters in the tens' place, and the number of living parents in the units' place.

Let us look at a typical calculation. Suppose one of the persons present has two brothers, three sisters, and two living parents.

First, he multiplies the number of his brothers by 2. $2 \times 2 = 4$.

Next, he adds 3. $3 + 4 = 7$.

Now, he multiplies the result by 5. $7 \times 5 = 35$.

Then, he adds the number of his sisters. $35 + 3 = 38$.

He multiplies the result by 10. $38 \times 10 = 380$.

He adds the number of his living parents. $380 + 2 = 382$.

Finally, he subtracts 150. $382 - 150 = 232$.

The figure in the hundreds' place in the number 232 is 2, which represents the number of his brothers.

The digit in the tens' place in the number 232 is 3, which represents the number of his sisters.

And the digit in the units' place in the number 232 is 2, which represents the number of his living parents.

Evidently, if the answer given you is 312, your friend has three brothers, one sister, and two living parents. Once you know his answer, you can't miss!

Blue-Streak Calculator

You tell your friend that you do not need a pocket calculator. You can do additions in your mind as fast as any machine. Let him put you to the test. If he wants to use a calculator to verify your solution, he is welcome to.

Ask him to write three four-digit numbers one below the other. Suppose he writes:

$$7214$$
$$8325$$
$$3742$$

Instantly you write below his numbers three of your own:

$$2786$$
$$1675$$
$$6258$$

Without a moment's pause, you write the sum of all six numbers: 30,000.

You may conceal this number, if you like, and wait patiently while your friend works it out for himself with pencil or calculator. Your answer, of course, agrees with his if he has done his arithmetic correctly.

The reason is that the numbers you added *had* to equal 30,000 when added to your friend's numbers, no matter what his numbers were. That is because you were careful, when you wrote your numbers, to match each of his numbers with one that adds to 10,000. $10,000 \times 3 = 30,000$. For example, his 7214 plus your 2786 is 10,000. The quick way to be sure of writing down the right number is to start with his digit in the thousands' place and add a digit to make 9, and keep matching 9's up to the units' digit, where you add a digit that will sum up to 10. For example, his first number, 7214, has a 7 in the thousands' place. You write down a 2 in that position in the number you add $(2 + 7 = 9)$. Corresponding to the 2 in the hundreds' place in his number, you write a 7 $(7 + 2 = 9)$ in the hundreds' place of your number. To the 1 in the tens' position in his number you add, in your number, an 8 $(8 + 1 = 9)$. But to the 4 in the units' place in his number you must add a 6 in the corresponding position in your number to make a sum of 10 $(6 + 4 = 10)$.

Note that the same procedure was followed in producing the other two numbers you added to his.

You can escalate this trick to five-digit, six-digit, and larger numbers, using more than three of each. You could also, in the example above, add any seventh number, say 9316, since this merely makes the sum 39,316 instead of 30,000. When you can reach solutions of such magnitude instantaneously, your friend will be convinced that you have an electronic computer in your brain.

It is a good idea not to try this trick too many times with the same person.

Subtraction Attraction

Besides being an addition magician, you have a subtraction attraction that is as fast as any electronic computer.

Ask a volunteer to subtract the square of a fairly large number—let us say, of 148—from the square of the next higher number—in this case, of 149. While he is laboriously writing the steps out or punching the keys on a pocket calculator, you are already reciting the answer. And yet *he* chose the numbers to be squared! Just to show that you can do the same with any number, you let another volunteer pick one, tell you and the rest of the group what it is, and start doing the arithmetic, while you jot down the answer instantly, to the bafflement of everybody.

You can repeat the trick as often as you like. It's really quite simple. All you do is add the two numbers. Their sum is the answer.

Suppose, for example, that we subtract $(148)^2$ from $(149)^2$. $(149)^2$ is 22,201, and $(148)^2$ is 21,904. $22,201 - 21,904 = 297$. This is the sum of 148 and 149.

Your reputation as a mathematical "whizzard" will really flourish when you perform the same trick with four-digit numbers. Take, for example, $(7562)^2 - (7561)^2$. $(7562)^2 = 57,183,844$. $(7561)^2 = 57,168,721$. $57,183,844 - 57,168,721 = 15,123 = 7562 + 7561$. These squares are so big that there is no room to list them in the table of Squares and Square Roots in the Appendix. Yet you can solve the problem in an instant mentally or jot down the answer in a flash if the numbers are too large for you to add easily in your head. You can escalate this as far as you want with the utmost confidence of always being ahead of your volunteer—provided he does not catch on to the trick!

The algebra, by the way, is: $(x + 1)^2 - x^2 = x^2 + 2x + 1 - x^2 = 2x + 1 = (x + 1) + x$.

Instant Arithmetic

You can boast to a friend that you have mastered the 15,873 multiplication table. You can, in fact, multiply by it instantly. If he doubts you, ask him to pick any number from 1 through 9, while you do the same. You write down your number—which is always 7—and his number and offer to multiply their product by 15,873 in three seconds. You can even take a bet on it, because you cannot lose—if you know the trick.

Suppose his number is 5. You just write it down six times: 555,555. That is the product of 35 (5 × 7) and 15,873. After your friend goes through the labor of verifying your answer he may wonder how you were able to get the correct answer in an instant.

207

Very simple. 15,873 multiplied by 7 (your number) is 111,111. Whatever single digit you multiply 111,111 by, from 1 through 9, you get a string of six digits identical to the multiplier, from 111,111 through 999,999. If your friend's number is 6, then the product (provided you always choose the number 7 as your multiplier of his number) must be 666,666. If his number is 8, the result will be 888,888. Just remember always to select the number 7 as yours.

This is not a trick to play more than once with the same friend — especially after winning a bet!

What Time Is It?

This is another mind-reading trick.

Ask a friend to think of any hour on the clock. Then have him mentally subtract it from 20 and remember the remainder. You now take your watch in hand and inform him that you are going to count 1, 2, 3 . . . , as you point at hours on the dial. When you count the number corresponding to the remainder that he was to remember, he is to stop you. But when he does, the hour you are pointing at is the hour he chose!

Here is how you do it. Say he thought of five o'clock. Then $20 - 5 = 15$, the remainder he has to remember. He will stop you at the fifteenth hour you point at, whatever it is. As you count from 1 to 7 you point at any hours you please, but you take care, when you count 8, to point at the twelve. Thereafter, you point to the hours in succession counterclockwise. That is, you count 9 and point at the eleven on the dial; you count 10 and point at the ten, etc. As you count 15, where your friend has to stop you, you are pointing at five o'clock, the hour he picked.

If you try this trick again with the same friend, have him subtract his hour from, say, 18 instead of 20. Since 18 is only 6 more than 12, make your sixth, rather than your eighth, count at twelve on the dial. In the first example, the eighth count was made at the twelve because 20 is 8 more than 12. If 22 is the number from which the subtraction is to be made, then the tenth count must be on the twelve on the dial because 22 is 10 more than 12.

By means of such variations, you can keep mystifying your friends indefinitely.

Number Seer

You announce to a group that you can fill a grid of eight columns by nine rows with numbers and see the seventy-two numbers in your mind with such clarity that, with your back turned or your eyes closed, you can call off all the numbers in any row. Indeed, if given a moment for deep concentration, you can even name the number in any box identified by row and column. This means, in effect, that you are undertaking to visualize the precise arrangement of seventy-two digits from 0 to 9 in the grid that you write at a blackboard or on a piece of graph paper, for all to see.

If your challenge is taken up, you prepare the grid illustrated. You then turn your back, or close your eyes, or even allow yourself to be blindfolded and invite anyone present to identify any row by number. Let us say that you are challenged to name the numbers in row 6. Instantly you reel off the numbers in order: 4, 2, 1, 4, 2, 8, 5, 6. And you do the same for any row. You can even let a little time elapse between demonstrations of your powers, just to show that your visual memory does not fade. And then, if you are asked to identify the number in row 9, column 6, you soon give the answer, 7.

The trick is in the pattern you use in putting the numbers in the grid. The key to the particular grid used here is the number 7. Whenever you multiply 7 by the number of a row—in this case, 6—you get the first two digits in that row, 4 and 2, from the product 42. Now if you multiply the units' digit, 2, in 42 by 7, you get 14, which gives you the next two digits in row 6: 1 and 4. Multiply the units' digit in 14 by 7: $4 \times 7 = 28$, and the next two digits in row 6 are 2 and 8. Finally, following the same procedure with the units' digit in this number, you get $8 \times 7 = 56$, which gives you the last two digits in row 6: 5 and 6. It's the same for every row. But note that in row 3, where 21 calls for $1 \times 7 = 7$, the next digits are 0 and 7.

Finding the exact number in any box identified by row and column requires a little more time, but the same procedure. For example, to find the number in row 9, column 6, mentally construct row 9 as above until you reach the second (units') digit of the third pair, which is in column 6. Thus, row 9 begins with $9 \times 7 = 63$. Multiply the 3 in 63 by 7 to give the next two digits: 2 and 1. You reach column 6 with the units' digit in $1 \times 7 = 07$, namely, 7.

After a while, you can make up a different grid, using a

different basic number instead of 7 as your multiplier, but the same system, to mystify your audience. If you let a confederate into your little secret, you can have him volunteer to write a grid for you, and you will memorize it at a glance. Just take a look to make sure that you understand it and that it contains no mistakes, and proceed as before. The effect is thoroughly baffling.

	1	2	3	4	5	6	7	8
1	0	7	4	9	6	3	2	1
2	1	4	2	8	5	6	4	2
3	2	1	0	7	4	9	6	3
4	2	8	5	6	4	2	1	4
5	3	5	3	5	3	5	3	5
6	4	2	1	4	2	8	5	6
7	4	9	6	3	2	1	0	7
8	5	6	4	2	1	4	2	8
9	6	3	2	1	0	7	4	9

ROW 6 → ←ROW 6

↑ COLUMN 6

Mathematical Clairvoyant

Give two people in a group sheets of paper and ask each one to write down and conceal from your view a two-digit number. Let us call one person A and the other B.

Have person A multiply his number by B's number. (Algebraically, this is ab.)

Now tell B to subtract 1 from his number ($b - 1$) and to subtract A's number from 100 ($100 - a$).

From here on B does all the mathematical work, but A can check its accuracy, and so can others if they wish.

B multiplies the two new numbers: $(b - 1)(100 - a)$.

Finally, B adds A's new number (ab) to the product: $(b - 1)(100 - a) + ab$.

You declare that you are a mathematical clairvoyant. You will tell each, A and B, his number if they will only tell you the answer to the problem you gave them.

Let us say they tell you that the answer, checked by both and perhaps others as well, is 8367. Instantly you tell A that his number is 67 and that B's is 84. And, of course, you are right, to their amazement.

The point is that $(b - 1)(100 - a) + ab = 100(b - 1) + a$. Then the "83" in 8367 (which is really 8300) is $(b - 1)$; so b is 84. And the 67 in 8367 is a, which is A's number, just as $b = 84$ is B's number. (Their chain of calculation was $67 \times 84 = 5628$; $b - 1 = 83$; $100 - a = 33$; $83 \times 33 = 2739$; $2739 + 5628 = 8367$.)

And that is all there is to it!

Whizzard

Tell a group that you are a "whizzard" (that is, a whiz and a wizard) with figures and shapes. Draw on the blackboard or on graph paper, for all to see, a square of thirty-six boxes, as illustrated. The rows are numbered upward from 1 through 6, and the columns similarly numbered, from left to right. Any box can then be numbered by column and row. Thus, the box in the upper left-hand corner, being in column 1, row 6, is given the number 16. The box in the lower right-hand corner is numbered 61, and the other boxes are numbered according to the same system.

After you leave the room or turn your back or are blindfolded, the group are to form any design of three adjacent boxes, whether horizontal boxes in a row, vertical boxes in a column, or a diagonal, an L- or U-shaped figure, or an L or U inverted, etc., anywhere on the grid. Three members of the group are assigned to remember one box number each corresponding to the three boxes forming the design. In the illustration, the group have picked a diagonal; A is to remember 24; B, 33; and C, 42.

You are then called into the room or your blindfold is removed and you undertake to reproduce the design and to place it exactly on the part of the grid where it belongs. All you need is a little cooperation from A, B, and C. Each is to take a piece of paper and do a little arithmetic.

First, each is to add 6 to his number.

Next, each is to multiply the sum by 2.

Now, each is to subtract 8.

Finally, each is to divide by 2.

After you get the three answers you fill in the design on

another grid of thirty-six boxes, placing it exactly where it belongs.

You can perform this surprising feat in record time simply by using a little algebra and plotting the points on the grid.

Call each volunteer's original number x.

Add 6. The result is $x + 6$.

Multiply by 2. The result is $2x + 12$.

Subtract 8. The result is $2x + 4$.

Divide by 2. The result is $x + 2$.

Say A called out 26; B, 35; and C, 44.

$x + 2 = 26$. Hence, A's number is 24. $x + 2 = 35$. Hence, B's is 33. $x + 2 = 44$. Hence, C's is 42.

You then plot column 2, row 4; column 3, row 3; and column 4, row 2; and you form the design selected by the group.

6	16	26	36	46	56	66
5	15	25	35	45	55	65
4	14	24	34	44	54	64
3	13	23	33	43	53	63
2	12	22	32	42	52	62
1	11	21	31	41	51	61
	1	2	3	4	5	6

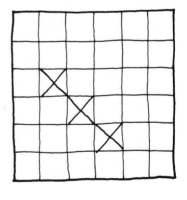

214

Naturally, you do not always have to specify the four operations as given. You can vary the operations, complicate them, or add more, so long as you remember how to use the algebraic answer to plot the design.

Figure Configurator

In this trick you ask anyone in your audience to draw on a blackboard or a pad three simple geometric shapes — say, a square, a circle, and a triangle, in that order — from left to right. You are told the configurations and their order. In each of these configurations, while you are blindfolded or your back is turned, any digit from 1 through 9 is to be placed. It does not matter whether the same digit is used in all three or one digit is repeated or three different digits are used. You will undertake to say what figure is in each configuration.

The members of the audience are to do a little arithmetic in accordance with your instructions. First, the number in the first configuration — in this case, the square — is to be doubled. To this 3 is to be added, and the sum is to be multiplied by 5. To the product is to be added the number in the next figure — here, the circle — and the sum is to be multiplied by 10. To this product is to be added the digit in the last shape (the triangle). All these calculations are silent; you see and hear nothing of them.

Now let someone in the audience call out a number from 1 through 10 and have him add it to the total reached above. As soon as you have been given the final sum, you name the digit in each geometric shape. You can make your performance more suspenseful by, say, first announcing the digit in the center (the circle), then – after pauses for concentration – revealing each of the other two.

You can further confuse the audience by seeming to approach the correct answer by a series of approximations, saying, for example, that the figure in the triangle appears to be odd and closer to 10 than the figure in the circle and in fact turns out (in this case, at least!) to be the sum of the digits in the square and the circle. Though all this is pure hocus-pocus, it can throw your audience off the scent.

And all you do is subtract, from the total given to you, 150 plus the number named in the last step.

The digits in the number you arrive at after this subtraction correspond exactly to those in the three geometric shapes and are arranged in the same order as they are.

Suppose, for example, that the digit in the square is a 2; in the circle, a 5; and in the triangle, a 7.

First, doubling the 2 gives 4. Next, add 3: $4 + 3 = 7$. Now, multiply by 5: $7 \times 5 = 35$. To this add the number in the circle (5): $35 + 5 = 40$. Multiply the sum by 10: $40 \times 10 = 400$.

Now, suppose that someone in the audience calls out the number 7. Add this to the previous total: $400 + 7 = 407$. This is the final result that you are given. You mentally subtract from this number $150 + 7$, (the number named in the last step): $407 - 157 = 257$. The 2 in this number is the 2 in the square; the 5 is the 5 in the circle; and the 7 is the 7 in the triangle.

Do you understand why this trick always works? Can you work it out by algebra? Essentially, you *undo* in a single step, in your mind, all the step-by-step calculations performed by the members of your audience, to get back to the original number.

Count-Down

Pile some buttons, coins, checkers, or tokens on a table. Tell a friend that you will play a game with him, taking turns at removing either one button or two buttons from the pile. The loser is the one who is left with the last button to remove. You let him begin. Every time he plays this game with you he is stuck with the last button.

How do you always force him to lose?

Just be sure to start with 22 or 19 or 16 or 13 or 10 or 7 or 4 buttons, that is, with a number that is 1 more than a multiple of 3 ("1 modulo 3"). If your opponent removes one button, you remove two; if he removes two, you remove one; so that the buttons decrease by threes until only one is left — and it's his turn. You cannot possibly lose.

Card Sensor

Tell a friend that you have the mysterious power to sense which card he has selected from a deck. He is to shuffle the deck thoroughly and divide it into two approximately equal piles.

He counts the cards in whichever pile he chooses. Let us say he counts twenty-three cards. He keeps this information to himself.

Next he is to add the two digits of the number – in his head. $2 + 3 = 5$. Once again he tells you nothing.

Finally, holding the cards with their faces turned toward him, he counts as far as the card representing the sum of the two digits, that is, the fifth card. Tell him to look at the card, concentrate on it, and remember it.

Now, keeping the cards exactly in the same order as before, he is to put his pile on top of the other pile and hand you the deck.

You take the deck, weigh it in each hand, smell it, and feel for the right card. You remove a card from the deck and show it to him. It is, of course, the card he selected.

What he does not know is that whatever card he selects must become the nineteenth card from the top of the deck. All you did was count down to the nineteenth card. That is all you need to do no matter how many times you perform this trick.

In fact, you can play it over the telephone. Have the person on the other end of the line read off the cards from the top of the deck, one at a time. Stop him when he reads the nineteenth card. It will always be the card he chose.

This trick depends on a simple arithmetical principle. For

each of the numbers from 20 through 29, the sum of the digits, when subtracted from the number, is always 18. For example, for 25, the sum of the digits is 7. $25 - 7 = 18$. For 28, the sum of the digits is 10. $28 - 10 = 18$. If a pack of cards is divided into approximately two equal piles, the number of cards in either pile is usually anywhere from 20 through 29. And if there are eighteen cards left over, the selected card must be the nineteenth card from the top of the deck when the cards are turned face down.

Add and Superadd

You announce to all present that you will undertake to state instantly the sum of a column of ten numbers if you are told only what the seventh number is.

Everyone who agrees to test your unusual arithmetical power is asked to write the digits from 1 through 10 in a column. Opposite the number 1 each may write any secret number he wishes, and another opposite the number 2. Now, opposite the number 3 he must write the sum of the numbers opposite 1 and 2. Opposite 4 he writes the sum of the numbers opposite 3 and 2; opposite 5, the sum of the numbers opposite 4 and 3; and so on through 10.

You now ask each one to add up his entire column of ten figures and then tell you what number he placed opposite 7. Once you know that, you instantly announce the sum of the entire column in each case, to the surprise of everyone. What is more, you perform this miraculous feat as many times as you are called upon to do so, without a mistake.

219

The secret consists in multiplying the number opposite 7 by 11. The result will always equal the sum of the column.

Here are two examples:

1.	3	17
2.	4	21
3.	$7 = 3 + 4$	$38 = 17 + 21$
4.	$11 = 7 + 4$	$59 = 21 + 38$
5.	$18 = 7 + 11$	$97 = 38 + 59$
6.	$29 = 11 + 18$	$156 = 59 + 97$
7.	$47 = 18 + 29$	$253 = 97 + 156$
8.	$76 = 29 + 47$	$409 = 156 + 253$
9.	$123 = 47 + 76$	$662 = 253 + 409$
10.	$\underline{199} = 76 + 123$	$\underline{1071} = 409 + 662$
	517	2783

7

Fun with Figures

MATHEMATICS IS a fascinating world, full of puzzling oddities, amazing curiosities, and bizarre and even sensational phenomena. Every step taken in the domain of numbers and shapes leads to surprising discoveries, amusing "novel-tease," intriguing mysteries, and entertaining marvels. Getting acquainted with the wealth of wonders to be found in mathematics enlivens the imagination, sharpens the perception of numerical and spatial relationships, stimulates inquisitiveness, and impels one to proceed further on one's own in an exciting and endless adventure with figures.

The problems in this chapter are more than just conundrums, posers, and riddles. Besides being fun in themselves, their solution involves practice in logical reasoning and computational skills as well as understanding of the processes of mathematical thought.

221

• Can you arrange the digits 0 through 9 so that they will add up to 1?

$$\frac{35}{70} + \frac{148}{296} = 1.$$

• Using the ten digits 0 through 9 without duplication, write a number equal to 100.

There are two ways of doing this: 78 3/6 + 21 45/90 = 100, and 50 1/2 + 49 38/76 = 100.

• From a number consisting of the ten digits 0 through 9 subtract another number consisting of the same ten digits and get a remainder consisting of the same ten digits.

$$\begin{array}{r} 9876543210 \\ - 1234567890 \\ \hline 8641975320 \end{array}$$

• Express the number 8 by using eight 8's. No fractions, please.

$8 \times (88{,}888)^{8-8} = 8$, since $(88{,}888)^0 = 1$.

• Express the number 2 with seven 2's and no other number.

$2 + 22/22 - 2/2 = 2.$

• Express 55, using only five 4's.

$44 + 44/4 = 55.$

• Express the number 1, using all nine digits from 1 through 9.

$1^{23456789} = 1$ is one way.

• Use four identical digits to express the number 12.

$11\ 1/1 = 12.$

• What are three ways of using five 9's to express the number 1?

$$.9\,\frac{99}{99} = 1.$$

$$.999\,\frac{9}{9} = 1.$$

$$(999)^{9-9} = 1.$$

• What are four ways of using four 5's to express the number 100?

$$(5+5) \times (5+5) = 100.$$
$$(5/.5) \times (5/.5) = 100.$$
$$55/.55 = 100.$$
$$\frac{(55-5)}{.5} = 100.$$

• Express the number 1000 with seven 9's.

$$999 + 99/99 = 1000.$$

• Take 45 from 45, and leave 45 as a remainder.

$$\begin{array}{r} 9+8+7+6+5+4+3+2+1 = 45 \\ -\ 1+2+3+4+5+6+7+8+9 = 45 \\ \hline 8+6+4+1+9+7+5+3+2 = 45 \end{array}$$

• Write the digits from 1 through 9 in such an order that the first three are a third of the last three, and the middle three are the first three taken from the last three.

219, 438, 657, because $219 = 657/3$ and $438 = 657 - 219$.

• Express the number 3 by using three 3's.

There are three ways of doing this:

$$3 + 3 - 3 = 3.$$
$$3 \times (3/3) = 3.$$
$$\sqrt[3]{3^3} = 3.$$

• The famous German mathematician Karl Gauss (1777–1855) was only seven years old when he surprised his teacher by solving in a few seconds the following problem, which he was given to keep him busy for a while: "What is the sum of all the numbers from 1 through 100?" The answer he gave, after only a moment's thought, was 5050. How could he be so sure so fast?

$100 + 1 = 101$; $99 + 2 = 101$; $98 + 3 = 101$, and so on. There are fifty pairs (100/2) of numbers between 1 and 100, including both, that add up, in every case, to 101. So just multiply 101 by 50 to get 5050.

• Take any three-digit number and multiply it by 143. Now repeat the digits in the three-digit number in the same order, making a six-digit number, and divide by 7. The result of your multiplication should be the same as the quotient you get by your division, no matter what the number.

For example, $651 \times 143 = 93{,}093$; $651{,}651 \div 7 = 93{,}093$.

Similarly, $892 \times 143 = 127{,}556$; $892{,}892 \div 7 = 127{,}556$.

• Can you make four 7's equal 100?

$$\frac{77}{.77} = 100.$$

• Can you make four 2's equal 7?

$$\frac{2/.2}{2} + 2 = 7.$$

• Can you make six 6's equal 37?

$$(6 \times 6) + \frac{66}{66} = 37.$$

• Can you take three letters from a word of four letters and leave five, without changing the word's meaning?

Take FIE from FIVE, leaving V, the Roman numeral for 5.

• What is the largest number that can be made with three digits?

9^{9^9}, which equals 9 raised to the 387,420,489th power — an enormous number.

• Take any three-digit number with the digits repeated, like 222. Divide the number by the sum of its digits. You will always get 37. See for yourself: $\frac{222}{2+2+2} = \frac{222}{6} = 37$. $\frac{999}{9+9+9} = \frac{999}{27} = 37$.

• Can you make three 7's equal 20?

$$\frac{7+7}{.7} = 20.$$

• Take any three-digit number — for example, 239 — and produce five other three-digit numbers by rearranging its digits — 293, 329, 392, 923, and 932. Add the six numbers and divide by 3. Then divide by the sum of the three digits in the original number. The result will always be 74. For example, $239 + 293 + 329 + 392 + 923 + 932 = 3108$; $3108 \div 3 = 1036$; $\frac{1036}{2+3+9} = \frac{1036}{14} = 74$.

• Make a six-digit number by repeating any three digits in the same order—for example, 562,562. With three simple divisions, can you bisect this number, that is, reduce it to the original three-digit number?

Divide by 7, 11, and 13. See for yourself: $562,562 \div 7 = 80,366$; $80,366 \div 11 = 7306$; $7306 \div 13 = 562$.   Hint: How much is $7 \times 11 \times 13$?

• How much is a third and a half-a-third of 5?

2 1/2 *or* 1 1/6, depending on whether you interpret "a third" as "a third of 5," that is, 1 2/3, or as simply 1/3.

• What number remains the same when it is turned upside down?

There are many such numbers: 1, 8, 88, 101, 1881, etc.

• Can you add 150 to a tree and make a garment?

CL + OAK = CLOAK. (CL is the Roman numeral for 150.)

• By adding 50 to a tree you can get a whip, and by adding 100 to the same tree you can get money. How?

Add L to ASH, and get LASH. Add C to ASH, and get CASH.

• By adding 1 to what you do when you play cards you can attain perfection. How?

Add I to DEAL, and get IDEAL.

• Can you add 5 to frozen water and get corruption?

Add V to ICE, and get VICE.

226

• Can you add 50 to the reward of work or service and get information and knowledge?

Add L to EARNING, and get LEARNING.

• Can you add 100 to a total and get a summons?

Add C to ALL, and get CALL.

• Can you add 1000 to what you do when you shoot a gun, and disfigure, mangle, or mutilate?

Add M to AIM, and get MAIM.

• Can you add 1000 to what you do when you give assistance, and get an assistant?

Add M to AID, and get MAID.

• Can you take 9 from 6, 10 from 9, and 50 from 40, ending with a remainder of 6?

$$
\begin{array}{ccc}
\text{SIX} & \text{IX} & \text{XL} \\
- \ \underline{\text{IX}} & \underline{\text{X}} & \underline{\text{L}} \\
\text{S} & \text{I} & \text{X}
\end{array}
$$

• Express the number 1000 with eight 8's.

$$\frac{8888 - 888}{8} = 1000.$$

• Can you double 421,052,631,578,947,368 instantly without using a calculator?

Shift the last digit to the front: 842,105,263,157,894,736.

• What is the square root of 123,456,787,654,321?

11,111,111. Check and see for yourself.

• The following numbers add the same when they are turned upside down and backward:

986	919
818	969
969	686
989	696
696	818
616	986
5074	5074

• Write any four consecutive odd numbers. Multiply the two end numbers, and then multiply the two middle ones. The difference between the products is always 8.

Try it and see. For example, with 1, 3, 5, and 7, $1 \times 7 = 7$, and $3 \times 5 = 15$. $15 - 7 = 8$. With 17, 19, 21, and 23, $17 \times 23 = 391$, and $19 \times 21 = 399$. $399 - 391 = 8$.

• How can you convert the number 617,283,945, which contains all the digits from 1 through 9, to 123,456,789?

Just divide by 5.

• Add 62 to any number from 51 to 100. Cross out the 1 in the hundreds place and add 1 to what is left. Then subtract this number from the original number. The remainder will always be 37.

Try it with 70 and see. Add 62. $62 + 70 = 132$. Remove the 1 from the hundreds place, leaving 32. Add the 1 to 32: $32 + 1 = 33$. Now subtract from the original number: $70 - 33 = 37$. You can't miss.

• What three consecutive numbers, when cubed and added, equal the cube of the next consecutive number?

3, 4, and 5. $3^3 + 4^3 + 5^3 = 6^3$. $3 \times 3 \times 3 = 27$. $4 \times 4 \times 4 = 64$. $5 \times 5 \times 5 = 125$. $27 + 64 + 125 = 216 = 6 \times 6 \times 6$.

• What two three-digit numbers, when multiplied, equal 111,111?

481 and 231, or 429 and 259.

• What two integers have 13 as their product?

13 and 1.

• How many half pints of milk are there in a half dozen?

6.

• A container holds exactly 8 quarts of milk and is filled to the top. Of two empty containers, one can hold 5 quarts, and the other, 3 quarts. By pouring the milk from one container to another, and without using any other measuring devices, can you divide the milk into equal parts of 4 quarts each?

In eight steps:

	8-quart jug	5-quart jug	3-quart jug
1	8	0	0
2	3	5	0
3	3	2	3
4	6	2	0
5	6	0	2
6	1	5	2
7	1	4	3
8	4	4	0

• With what five weights can you weigh anything from half a pound up to sixty pounds, either in whole pounds or half pounds?

1/2 pound, 1 1/2 pounds, 4 1/2 pounds, 13 1/2 pounds, and 40 1/2 pounds.

For example, if goods weighing 3 pounds are placed on the same side of the scale as a weight of 1 1/2 pounds, the counterbalance will be a weight of 4 1/2 pounds. With a counterbalance of 13 1/2 pounds, goods placed on the other side of the scale can be shown to weigh 7 pounds when added to weights of 4 1/2 pounds, 1 1/2 pounds, and 1/2 pound. Similarly, with a weight of 40 1/2 pounds on one side of the scale, it will balance if you place on the other side weights of 13 1/2 pounds, 1 1/2 pounds, 1/2 pound, and goods that weigh 25 pounds.

• What seven weights would you need to determine any weight of a whole number of pounds from 1 pound to 127 pounds?

You would need weights of 1, 2, 4, 8, 16, 32, and 64 pounds.

For example, to counterbalance a weight of 65 pounds, you would place on the other side of the scale weights of 64 pounds and 1 pound. To weigh something having a weight of 53 pounds, you counterbalance it with weights of 32, 16, 4, and 1 pounds. To determine that something weighs 91 pounds, place on the other side of the scale the counterbalancing weights of 1, 2, 8, 16, and 64 pounds.

• If you subtract 10 from a number, multiply by 3, find the square root, and subtract 18, you will have nothing left. What is the number?

You solve this problem by starting from 0 and working backward: $0 + 18 = 18$; $18^2 = 324$; $324 \div 3 = 108$; $108 + 10 = 118$. Check: $118 - 10 = 108$. $108 \times 3 = 324$. $\sqrt{324} = 18$. $18 - 18 = 0$.

• How can you tell whether a large number is divisible by 11?

Add the first, third, fifth, and succeeding odd digits. Then add the second, fourth, sixth, and succeeding even digits. If the difference between the two sums is 0 or divisible by 11, then the number is divisible by 11. For example, take the number 1,075,624. The first digit is 1; the third, 7; the fifth, 6; and the seventh, 4. $1 + 7 + 6 + 4 = 18$. The second digit is 0; the fourth, 5; the sixth, 2. $0 + 5 + 2 = 7$. $18 - 7 = 11$, which is divisible by 11. Therefore, 1,075,624 is divisible by 11. (The quotient is 97,784.)

• If a man can buy puppies for $5 each, kittens for $3 each, and rabbits for 50¢ each, and if he bought 100 animals for $100, how many of each did he get?

Either he got 10 puppies ($50), 2 kittens ($6), and 88 rabbits ($44), or he purchased 5 puppies ($25), 11 kittens ($33), and 84 rabbits ($42). $10 + 2 + 88 = 100 = 5 + 11 + 84$. $\$50.00 + \$6.00 + \$44.00 = \$100.00 = \$25.00 + \$33.00 + \$42.00$.

• Starting with $16.00, you bet half on the toss of a coin. Win or lose, you bet half of what you then have on a second toss. You continue this process through six tosses. If you win three tosses and lose three, how much money do you have left?

You will have lost $9.25. Therefore, you will end with $6.75. It makes no difference in what order you win or lose as long as you win half and lose half of the tosses.

• What percent profit does a bank make if it borrows money at 4 percent and lends the same money at 6 percent?

50 percent. It pays $40 interest per $1000 while collecting $60. The $20 difference is 50 percent of the $40 invested by the bank.

• If it takes a church clock 12 seconds to strike four, how long does it take to strike eight?

28 seconds. Between the first and the fourth strikes there are three intervals of silence that take 12 seconds, or 4 seconds per interval. Four more strikes means four more intervals, which adds 16 seconds to the original 12 seconds.

• In a family in which each son has twice as many sisters as he has brothers, and each daughter has as many brothers as she has sisters, how many sons and daughters are there?

There are three sons and four daughters. Each son has two brothers and twice as many ($2 \times 2 = 4$) sisters. Each daughter has three brothers and an equal number of sisters.

• A used-car dealer told a friend that he had had a bad day. He had sold two cars for $750 each, one for 25 percent profit and the other for a 25 percent loss. His friend said it wasn't such a bad day, because his profit and loss canceled each other out. Who was right?

The dealer was right. If he made a 25 percent profit on a car that he sold for $750, it must have cost him $600. But if he lost 25 percent on a car that he sold for $750, it must have cost him $1000. So the two cars cost him $1600 and sold for only $1500. Hence, he suffered a $100 loss that day.

• You have nine ball bearings, eight of equal weight and one heavier. Using a balance scale, can you locate the heavy ball in two weighings?

Put three balls on each pan. The heavy ball is in the set that sinks; or if neither sinks it is in the set left off the balance. Whichever set of three the heavy ball is in, put one ball on each pan. The heavy ball is the one that sinks; or if neither sinks it is the one left off the balance.

• The students in a mathematics class want to know how old their teacher is. He says that he has lived 1/4 of his life as a boy, 1/5 as a young man, 1/3 as a mature man, and 13 years to his present age. How old is he?

60. Let x be his age. Then $\frac{x}{4} + \frac{x}{5} + \frac{x}{3} + 13 = x$. Solving, $\frac{15x}{60} + \frac{12x}{60} + \frac{20x}{60} + 13 = x$. $\frac{47x}{60} + 13 = x$. $13 = x - \frac{47x}{60}$. $13 = \frac{60x}{60} - \frac{47x}{60} = \frac{x(60-47)}{60} = \frac{13x}{60}$. Dividing each side of the equation by 13, we get: $1 = \frac{x}{60}$. $x = 60$. Check: 15 years as a boy, 12 as a young man, 20 as a mature man, and 13 more; total 60.

• The sum of the digits in the eight-digit number 87,654,321 is 36.

Now reverse the order of the digits: 12,345,678. Naturally, the sum of the digits is still 36.

When you subtract the smaller number from the greater, is the sum of the digits 0?

$$
\begin{array}{r}
87,654,321 \\
- \ 12,345,678 \\
\hline
75,308,643
\end{array}
$$

No, it's 36!

• Can you make three 6's equal 7?

$$6 \ 6/6 = 7.$$

• Can you make three 4's equal 11?

$$\frac{44}{4} = 11.$$

- Can you make three 3's equal 24?

$$3^3 - 3 = 24.$$

- Can you make three 5's equal 10?

$$(\sqrt{5} \times \sqrt{5}) + 5 = 10.$$

- Can you make three 5's equal 5?

$$5 + 5 - 5 \text{ or } \frac{5}{5} \times 5 \text{ or } \sqrt[5]{5^5} = 5.$$

- Can you make five 3's equal 42?

$$33 + 3 + 3 + 3 = 42.$$

- Write three different numbers at the angles of a triangle —
say, 3, 6, and 5, as illustrated.

At the center of each side of the triangle write the sum of
the numbers at either end of it: 9, 8, and 11.

Now, draw a line from each angle to the center of the opposite side. You will find that invariably, if you add the
numbers at the extremities of the lines just drawn, their
sums will be the same — in this case, 14 — which can be
written at the point of their intersection, as in the illustration.

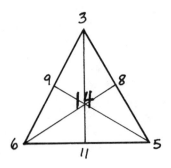

234

It does not matter what numbers you use – it always works. Try it with 1, 2, and 4. In that case the common sum will be 7. If you use 27, 30, and 42, the common sum will be 99.

• Using the digits from 1 through 9 once each, can you form each of the following fractions: 1/2, 1/3, 1/4, 1/5, 1/6, 1/7, 1/8, and 1/9?

$$1/2 = \frac{9267}{18,534} \qquad 1/6 = \frac{2943}{17,658}$$

$$1/3 = \frac{5823}{17,469} \qquad 1/7 = \frac{5274}{36,918}$$

$$1/4 = \frac{7956}{31,824} \qquad 1/8 = \frac{9321}{74,568}$$

$$1/5 = \frac{2973}{14,865} \qquad 1/9 = \frac{8361}{75,249}$$

• Can you make four 4's equal 3, 9, 36, and 45?

$$\frac{4+4+4}{4} = 3. \qquad\qquad 4+4+\frac{4}{4} = 9.$$

$$44 + \frac{4}{4} = 45. \qquad 4(4+4)+4 = 36.$$

• Can you make five 1's equal 14?

$$11 + 1 + 1 + 1 = 14.$$

• Can you use all the digits from 1 through 9 just once to equal 45?

There are two ways of doing so: $\dfrac{5 \times 8 \times 9 \times (7+2)}{1 \times 3 \times 4 \times 6} = 45.$

$7^2 - \dfrac{5 \times 8 \times 9}{3 \times 4 \times 6} + 1 = 45.$

• Using the digits from 1 through 9 just once in each fraction can you form two fractions that, when multiplied, will be equal to another fraction formed with the same nine digits differently arranged?

$$\frac{18,534}{9,267} \times \frac{17,469}{5,823} = \frac{34,182}{5,697}.$$

You will find that this equation is equivalent to: $2 \times 3 = 6$.

• Can you represent the number 9 by a fraction containing all the digits from 0 through 9, used once each?

There are six such fractions: $\frac{97,524}{10,836}$, $\frac{95,823}{10,647}$, $\frac{95,742}{10,638}$, $\frac{75,249}{08,361}$, $\frac{58,239}{06,471}$, and $\frac{57,429}{06,381}$. But the last three are debatable (look at the denominators).

• Here are some oddities involving squares. Both can be continued indefinitely.

$4^2 =$	16	$7^2 =$	49
$34^2 =$	1156	$67^2 =$	4489
$334^2 =$	111,556	$667^2 =$	444,889
$3334^2 =$	11,115,556	$6667^2 =$	44,448,889

• If you have a pocket calculator, you can get it to answer some unusual questions.

"What is the best business to be in nowadays?"

Just use your calculator to multiply 473,849 by 15. Don't bother reading the product. Just turn the calculator upside down and read: "SELL OIL."

Again, if 14 million motorists in the United States are on the highways every day, plus 215,000 drivers lined up at gas stations, while 469 Saudi Arabian sheiks are having

236

a business meeting, how many people are involved in five days, and who wins in this situation? Add all the people, and multiply by 5 to get the first answer; for the second, turn the calculator upside down and read: "SHELL OIL."

• A man bought a hat for $5 and handed the storekeeper a $50 bill to pay for it. Unable to make change, the merchant sent the bill to the grocer next door, got it changed, and gave the man who bought the hat $45. After the customer had left the store, the grocer, discovering the bill to be counterfeit, returned it to the hatter, who gave the grocer $50. How much did the owner of the hat store lose by the transactions?

He lost $50−$5 for the value of the hat plus $45 in cash which he gave the customer. The $50 the hatter gave the grocer was simply a return of the $50 the hatter had received from the grocer earlier in exchange for the counterfeit bill.

• Each day Tom gets to work on time his father gives him $10, but he takes away $15 each day Tom is late. After twenty days Tom has neither gained nor lost any money. How many days did he get to work on time?

12 days. Let x be the number of days Tom got to work on time. Then $20 - x$ is the number of days he was late. $10x - (20 - x)(15) = 0.$ $10x - (300 - 15x) = 0.$ $10x - 300 + 15x = 0.$ $25x = 300.$ $x = \dfrac{300}{25} = 12.$ Check: Tom got $120 and gave back $120.

• There are a hundred gloves in a drawer, half left-handed and half right-handed. Reaching into the drawer without looking, how many gloves must you take out to guarantee that you will have a pair—that is, at least one left-handed and one right-handed glove?

Fifty-one. The first fifty might be all for one hand, but no more.

• There are an unlimited number of five kinds of candies in a bag. They are all the same shape, so you can't tell which kind is which by feeling them. How many candies must you draw from the bag, without looking, to be sure of getting three of some kind?

Eleven. If you drew ten, you might have two of each of the five kinds. But the eleventh candy would have to be the third of some kind.

• A farmer's chickens and sheep have 24 heads and 76 feet. How many chickens does he have, and how many sheep?

Using the obvious letters, $c + s = 24$, and $2c + 4s = 76$. Substitute $(24 - s)$ for c in the second equation and solve for s, which is 14 sheep, while c is $24 - s = 10$ chickens.

• In what year did the twentieth century begin?

1901. There are 100 years in a century. There was no year 0 (A.D. 1 was the next year after 1 B.C.). The first century began with the year 1 and ended with the year 100. Therefore, the nineteenth century began with the year 1801 and ended with the year 1900.

• Make a number using the digits from 1 through 9 in sequence. Now make another number, reversing the order of the digits in the first number. Add the two numbers and add one. The result is a number consisting exclusively of 1's.

238

See for yourself:

$$123,456,789$$
$$987,654,321$$
$$\underline{1}$$
$$1,111,111,111$$

• Make a number using the digits from 1 through 9 in sequence. Multiply by 8, and add 9. The sum will be a number consisting of all the digits in the first number in reverse order.

See for yourself:

$$123,456,789$$
$$\underline{\times\ 8}$$
$$987,654,312$$
$$\underline{+\ 9}$$
$$987,654,321$$

• Multiply any two-digit number by 100, and then subtract the original number from the product. The remainder will always be a number whose digits total 18.

Examples: $29 \times 100 = 2900$; $2900 - 29 = 2871$; $2 + 8 + 7 + 1 = 18$. $35 \times 100 = 3500$; $3500 - 35 = 3465$; $3 + 4 + 6 + 5 = 18$.

• How many 9's are there between 1 and 100?

Twenty: $9, 19, \ldots, 89; 90, 91, \ldots, 98$; and two 9's in 99.

• Can you make four 9's equal 100?

There are two ways:

$$99\ 9/9 = 100.$$

$$\frac{99}{.99} = 100.$$

• Can you make six 9's equal 100?

$$[(9 \times 9) + 9] + 9\frac{9}{9} = 100.$$

• An oddity of the number 142,857 is that if it is multiplied by any number from 2 through 6, the product consists of the same digits in the same order, but always beginning with a different digit. See for yourself: $142{,}857 \times 2 = 285{,}714$; $142{,}857 \times 3 = 428{,}571$; $142{,}857 \times 4 = 571{,}428$; $142{,}857 \times 5 = 714{,}285$; and $142{,}857 \times 6 = 857{,}142$. But if you multiply 142,857 by 7, you get 999,999.

• Can you arrange eight 4's to total 500?

$$4 + 4 + 4 + 44 + 444 = 500.$$

• Can you divide a square into four equal triangles with two lines?

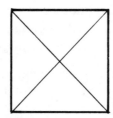

• Can you arrange nine matches, all the same size, to form three equal squares and two equal triangles?

Yes, but only in three dimensions. It's a prism:

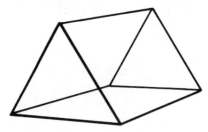

240

• One of the most mysterious and intriguing numbers is 9. In ancient lore, in the Hebrew cabala, in Sanskrit writings, in Christian tradition, and in history and literature, the number 9 and its oddities play an important role.

In Greek mythology there were nine muses or goddesses, each of whom had her domain—in epic song, lyric song, history, comedy, tragedy, dancing, love poetry, sacred song, and astronomy.

The number 9 is the highest number that can be expressed in a single digit.

Ancient tradition ascribed nine "planets" relating to life on earth (from Mercury through Saturn plus the sun and moon), nine heavens, nine orders of angels, and nine regions of hell.

There are nine months of pregnancy, possession is nine points of the law, a cat has nine lives, a cat-o'-nine-tails is used for punishment, and leases are often made for ninety-nine years.

Here are a few of the many mathematical oddities connected with the number 9:

Arrange the digits from 1 through 8 in a row. Work inward, removing pairs of end digits. Each pair of digits will add up to 9:

$$1 \; 2 \; 3 \; 4 \; 5 \; 6 \; 7 \; 8$$

$$1 + 8 = 9 \qquad 3 + 6 = 9$$
$$2 + 7 = 9 \qquad 4 + 5 = 9$$

Take any number from 1 through 9 and multiply it by 9. The digits in the product always add up to 9. See for yourself:

$1 \times 9 = 9.$
$2 \times 9 = 18.$ $1 + 8 = 9$
$3 \times 9 = 27.$ $2 + 7 = 9$
$4 \times 9 = 36.$ $3 + 6 = 9$
$5 \times 9 = 45.$ $4 + 5 = 9$
$6 \times 9 = 54.$ $5 + 4 = 9$
$7 \times 9 = 63.$ $6 + 3 = 9$
$8 \times 9 = 72.$ $7 + 2 = 9$
$9 \times 9 = 81.$ $8 + 1 = 9$

In fact, if you multiply *any* number by 9, the sum of the digits in the product will equal either 9 or a multiple of 9. For example, $395 \times 9 = 3555$, and $3 + 5 + 5 + 5 = 18 = 9 \times 2$, while $864 \times 9 = 7776$, and $7 + 7 + 7 + 6 = 27 = 9 \times 3$.

The following table of products and sums demonstrates another curious feature of the number 9:

$$(0 \times 9) + 1 = 1$$
$$(1 \times 9) + 2 = 11$$
$$(2 \times 9) + 3 = 21$$
$$(3 \times 9) + 4 = 31$$
$$(4 \times 9) + 5 = 41$$
$$(5 \times 9) + 6 = 51$$
$$(6 \times 9) + 7 = 61$$
$$(7 \times 9) + 8 = 71$$
$$(8 \times 9) + 9 = 81$$
$$(9 \times 9) + 10 = 91$$
$$(10 \times 9) + 11 = 101$$
$$(11 \times 9) + 12 = 111$$
$$(12 \times 9) + 13 = 121$$
$$(13 \times 9) + 14 = 131$$

And here's another list (of products):

$$999{,}999 \times 2 = 1{,}999{,}998$$
$$999{,}999 \times 3 = 2{,}999{,}997$$
$$999{,}999 \times 4 = 3{,}999{,}996$$
$$999{,}999 \times 5 = 4{,}999{,}995$$
$$999{,}999 \times 6 = 5{,}999{,}994$$
$$999{,}999 \times 7 = 6{,}999{,}993$$
$$999{,}999 \times 8 = 7{,}999{,}992$$
$$999{,}999 \times 9 = 8{,}999{,}991$$

Note that the first and last digits in each product add up to 9.

Can you add the digits from 1 through 9 to total 99,999?

$$\begin{array}{r} 1{,}234 \\ \underline{98{,}765} \\ 99{,}999 \end{array}$$

One more table of nine-ish products:

$$9 \times 9 = 81$$
$$99 \times 99 = 9801$$
$$999 \times 999 = 998{,}001$$
$$9999 \times 9999 = 99{,}980{,}001$$
$$99{,}999 \times 99{,}999 = 9{,}999{,}800{,}001$$

Form a figure consisting of any number of digits. Multiply it by 10 (that is, add a zero). Subtract the original number from the product. The remainder will always be a number whose digits total 9 or a multiple of 9. See for yourself:

$$65{,}341 \times 10 = 653{,}410. \quad 653{,}410 - 65{,}341 = 588{,}069.$$
$$5 + 8 + 8 + 0 + 6 + 9 = 36 = 9 \times 4.$$
$$7265 \times 10 = 72{,}650. \quad 72{,}650 - 7265 = 65{,}385.$$
$$6 + 5 + 3 + 8 + 5 = 27 = 9 \times 3.$$

Add the digits in a two-digit number and subtract the sum from the original number. The digits in the answer will always add up to 9. See for yourself:

39. $3 + 9 = 12$. $39 - 12 = 27$. $2 + 7 = 9$.
52. $5 + 2 = 7$. $52 - 7 = 45$. $4 + 5 = 9$.

And here is still another oddity of the number 9.

$12{,}345{,}679 \times 999{,}999{,}999 = 12{,}345{,}678{,}987{,}654{,}321$. You can check this with a calculator.

• A boy went to bed at eight o'clock in the evening and set the alarm, because he had to get up at nine in the morning. How many hours of sleep did he get?

Just one. At nine o'clock the same night, the alarm went off.

• If the earth's surface were absolutely smooth so that you could tie a ribbon tightly all around it, how much would you have to add to the ribbon's length so that it would be exactly 1 inch above the surface of the earth all the way around.

Only 2π, or about 6.28, inches.

• From a keg containing 20 gallons of wine a pint is drawn and poured into a pitcher containing 2 quarts of water. The wine is mixed thoroughly with the water in the pitcher, and a pint of the mixture is poured back into the keg. Will there then be more or less water in the keg than there is wine in the pitcher?

The amount of water in the keg will be equal to the amount of wine in the pitcher. It does not matter what the capacities of each may be. As long as each contains the same volume of liquid after the mixing of the wine and water as before, equal amounts of wine and water must have changed places.

244

• On returning from the dairy, three cooperative farmers have seven full pails of milk and seven pails half full. They also have seven empty pails. How can they divide what they have, without pouring any milk from one pail to another, so that each farmer will get the same amount of milk and the same number of pails to take home?

There are two possible arrangements. Two farmers can each get two full pails, three half-full pails, and two empty pails, making seven pails in all, with 3 1/2 pails of milk, while the third gets three full pails, one half-full pail, and three empty pails. Or two farmers can each get three full pails, one half-full pail, and three empty pails, while the third gets one full pail, five half-full pails, and one empty pail. Either way, each farmer gets seven pails and 3 1/2 pails of milk.

• A horse travels half a trip at 12 miles per hour and half at 4 miles per hour. What is its average speed?

6 (not 8) miles per hour. By definition, speed, r (rate in miles per hour), is distance, d (in miles), divided by time, t (in hours). For convenience, call half the distance x; in the first half of the trip, since $t = d/r$, $t = x/12$, and in the second half, $t = x/4$. Since $d = 2x$, the speed of the whole trip is $r = \dfrac{2x}{\dfrac{x}{12} + \dfrac{x}{4}} = \dfrac{2x}{\dfrac{4x}{(12)}} = \dfrac{24x}{4x} = 6.$

• If a hundred houses are lined up in rows of ten and numbered consecutively, and if houses no. 20 through no. 30 burn down, how many houses are left?

Eighty-nine.

• Walking up the street to the top of a hill, you count fifty windows on your right. You turn around and descend while counting fifty windows on your left. How many windows are included in your count?

Fifty. You counted the same fifty windows a second time as you descended.

• How would you go about determining the exact weight of just one page in the telephone book?

Weigh the book on a bathroom scale, and divide by the number of pages in the book. (Note: "pages 1–1000" means only 500 pages to weigh.)

• Of what two numbers is the sum equal to the product?

Elementary algebra shows that there are uncountably many such pairs of numbers. Let x and y be the two numbers; then $xy = x + y$, and $x = \dfrac{y}{y - 1}$. Therefore, for every value of y (except 1), there is one value of x, e.g., $y = -1, x = 1/2$; $y = 0, x = 0$; $y = 3/4, x = -3$; $y = 3, x = 3/2$; . . .

• Starting at twelve o'clock, how many times a day would the hands of a clock meet if the minute hand moved backward? (The hour hand moves forward, as usual.)

Twenty-five times.

• Seven customers eat at a restaurant regularly. One comes every day, one every other day, one every third day, one every fourth day, one every fifth day, one every sixth day, and one every seventh day. How often are they all there at once?

The least common multiple of 1, 2, 3, 4, 5, 6, and 7 is 420. Every 420 days.

246

• Can you make a magic square in which the *product* of each row, column, and diagonal is 1000?

50	1	20
4	10	25
5	100	2

• A thousand and one
  And a sixth part of twenty:
  Some may have none,
  But others have plenty.
  What is it?

Money. 1000 = M (Roman numeral). One = ONE, and Y = 1/6 of the word "twenty."

• Take part of a foot,
  And with judgment transpose;
  You'll find that you have it
  Just under your nose.
  What is it?

Chin. Transpose the first half of "inch" to the end.

• I may be half of ten;
  I may be nearly nine.
  If eight contains me,
  Most of six is mine.
  A third of one, a fourth of four . . .
  What am I?

0 (zero). 0 is the second half, reading from left to right, of 10. If you remove the stem from 9, what is left is 0. There are two 0's (circles) in 8, and 0 plus a stem makes 6. A zero looks like the O which is a third of the letters in ONE and a fourth of those in FOUR.

8

Making Cents of Dollars

AN INDISPENSABLE MEANS of learning the serious business of making sense of money is learning to make cents of dollars.

Through handling coins and currency of various denominations and making change, youngsters become familiar with fractional and decimal relations, conversions, ratios, and equivalences. At the same time, they can apply their arithmetical skills to practical affairs.

Pondering and solving the problems in this chapter is a pleasant and diverting way to become acquainted with the peculiarities of our monetary system, and an easy way to become accustomed to the quick and accurate computations needed in shopping, spending one's allowance, preparing a budget, and engaging in commercial transactions.

• I have two coins that total 55¢. One of them is not a nickel. What are the two coins?

A 50¢ piece and a nickel. (The coin that is not a nickel is the 50¢ piece.)

• I have six coins that total $1.15. Yet I cannot make change of a dollar or of a half dollar or of a quarter or of a dime or of a nickel. What six coins do I have?

A half dollar, a quarter, and four dimes.

• A man paid out $63 in six bills, none of them $1 bills. How did he do it?

He used one $50 bill, one $5 bill, and four $2 bills.

• A lady spent 35¢ for some stationery and handed the storekeeper a half dollar.

After looking in his till he said, "I'm sorry, but I can't change a half dollar. I can change a dollar, though."

What coins, besides pennies, did he have in his till?

A quarter and four dimes.

• What is the smallest number of coins that can total 99¢?

Eight (a half dollar, a quarter, two dimes, and four pennies).

• John had 23 coins, including dimes, quarters, and half dollars. After he changed the dimes into pennies, the quarters into nickels, and the half dollars into quarters, he had 110 coins. How many coins of each denomination did he start with?

Five dimes, eight quarters, and ten half dollars; or two dimes, sixteen quarters, and five half dollars.

250

• Can you change a dollar into fifty coins?

Forty pennies, eight nickels, and two dimes; or one quarter, two dimes, two nickels, and forty-five pennies.

• How many different combinations of coins (other than pennies) add up to 30¢?

Five (three dimes, six nickels, a quarter and a nickel, four nickels and a dime, and two nickels and two dimes).

• A charitable gentleman, on meeting a beggar, found he had only coins in his purse. He gave the beggar half of what he had and one penny more. On meeting another beggar, he gave him half of what he had left and one penny more. On meeting a third beggar, he gave him half of what he had left and four pennies more. He had four pennies left. How much did he have to begin with?

70¢. After the first beggar took 36¢, he had 34¢. After the second beggar took 18¢, he had 16¢. After the third beggar took 12¢, he had 4¢.

The way to solve this kind of problem without a lot of parentheses within brackets within braces is to work backward. Let c be how much money the gentleman had after giving to the *second* beggar. Half c is four pennies and four pennies more; so $c = 16$¢. Let $b =$ his money after giving to the first beggar. Half b is 16¢ and a penny; so $b = 34$¢. Let $a =$ the money he began with. Half a is 34¢ and a penny; so $a = 70$¢.

• A boy will get an allowance of either $1 or $10, depending on how he solves the following problem. His father gives him ten $1 bills and ten $10 bills. He is to divide the twenty bills into two sets, with any number of bills in each set and the denominations mixed in any way he pleases.

After he prepares the two sets he is blindfolded, and each set is placed in a separate hat. He must pick a hat blind and take one bill out of it, which will be his allowance. How should he construct the sets to get the best odds of getting an allowance of $10?

He should place one $10 bill in one set and all nineteen other bills in the second set. Now he is sure of picking a $10 bill if he chooses the hat with the first set, and he has almost an even chance of getting a $10 bill if he chooses the hat with the second set. Thus he has almost a 75 percent chance (actually, 73.68 percent) of getting a $10 bill, compared with, say, 50 percent if he makes any arrangement of two sets of ten bills each.

• Jack said to Bill, "Give me a dollar, and I'll have as much money as you have."

Bill said to Jack, "No, you give me a dollar, and I'll have twice as much as you have."

How much money does each have?

Jack has $5 and Bill has $7. If Bill gives Jack $1 they will both have $6; if Jack gives Bill $1, Jack will have only $4 to Bill's $8.

You can solve this problem with simultaneous equations: let x be how much Jack has and y how much Bill has. Then $x + 1 = y - 1$ and $y + 1 = 2(x - 1)$. To solve, substitute $y = x + 2$ from the first equation for y in the second equation, so that $x + 2 + 1 = 2(x - 1)$. Then $x + 3 = 2x - 2$. $x = 5$. $y = 7$.

• A family pools its funds to take a trip. Mother contributes one-third of the money needed; James, one-eighth; and Mary, one-tenth. Dad makes up the difference of $53. How much do they plan to spend on the trip?

$120. Mother contributes $40; James, $15; Mary, $12; and Dad, $53, which adds up to $120.

If x is how much the family plans to spend, $\dfrac{x}{3} + \dfrac{x}{8} + \dfrac{x}{10} + 53 = x$. Multiply both sides of the equation by 120. $53x = 53(120)$. $x = 120$.

• After paying 31¢ for three oranges and four lemons, Jane changed her mind. She went back to the fruit stand, returned a lemon, and took another orange instead.

"That will cost you another cent," said the fruit dealer.

How much does an orange cost? A lemon?

Oranges are 5¢ and lemons are 4¢. Three oranges and four lemons is $15 + 16 = 31$¢; four oranges and three lemons is $20 + 12 = 32$¢.

The simultaneous equations are $3o + 4l = 31$ and $4o + 3l = 32$.

• John bought a shirt, three ties, and seven handkerchiefs, paying $14. Jim liked what John bought so much that he bought the same styles: one shirt, four ties, and ten handkerchiefs, paying $17. How much is one shirt, one tie, and one handkerchief? And how much would it have cost to purchase two shirts, three ties, and five handkerchiefs?

You cannot find the cost of one shirt, or one tie, or one handkerchief, because the information given leads to only two simultaneous equations with three unknowns:

$$s + 3t + 7h = 14$$
$$s + 4t + 10h = 17$$

We will eliminate s and t, and hope that h will eliminate itself in the combinations asked for. (If not, the questions cannot be answered.)

Subtract the first equation from the second, obtaining $t = 3 - 3h$. Substitute $(3 - 3h)$ for t in the first equation, obtaining $s = 5 + 2h$. Finally, substitute $(3 - 3h)$ for t and $(t + 2h)$ for s in the combinations asked for:

$$s + t + h = 5 + 2h + 3 - 3h + h$$
$$2s + 3t + 5h = 10 + 4h + 9 - 9h + 5h$$

The h's do cancel out on the right side, leaving the answers: $8 and $19, respectively.

• You have to pay $103 for a painting. Can you do it with eight bills? (Not five $20 bills and three $1 bills.)

Use one $50 bill, two $20 bills, one $5 bill, and four $2 bills.

• Can you change $15 into a hundred coins? (No silver dollars, please.)

40 pennies, 14 nickels, 14 dimes, 14 quarters, and 18 half dollars is 40¢ + 70¢ + $1.40 + $3.50 + $9 = $15.

9

Shaping Figures
and Figuring Shapes

AN ESSENTIAL MATHEMATICAL SKILL is the
ability to recognize, visualize, and construct, either on
paper or at a blackboard, with the aid of a straightedge and
compass, a variety of geometric figures — triangles, squares
and other parallelograms, trapezoids, other polygons,
circles, and ellipses.

Familiarity with these shapes promotes an understanding
of the spatial and mathematical relations among them and
strengthens the visual imagination.

The problems in this chapter call for such activities as
folding, cutting, overlaying, and drawing geometric configurations. Some of them require such easily accessible
materials as coins, toothpicks, or matches. They offer
interesting challenges that heighten insight into angularities
and curvatures and stimulate an appreciation of limits,
contours, circumscriptions, contiguities, congruences, re-

flections, symmetries, and proportions in geometric constructions — and, sometimes, shortcuts for making shapes.

• Here is a quick way to make a square.

Start with a rectangular sheet of paper. Fold it so that one of the right angles is bisected, as by NK in the illustration. Then M should lie over S, showing that SK makes a right angle with NS, and a fold along SK will produce the square $NSKM$.

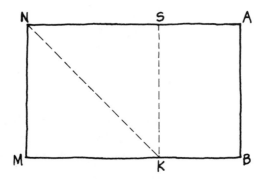

• Starting with any square, you can make new squares, one after another, each having exactly one-half the area of its parent.

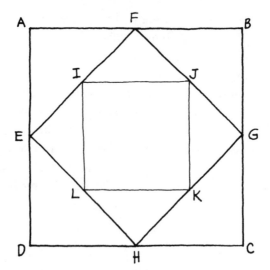

Fold each corner of the square to the center along the lines *EF*, *FG*, *GH*, and *HE* in the illustration. These lines define a new square, *EFGH*, with half the area of *ABCD*. By the same process you form the third square, *IJKL*, with half the area of *EFGH*, and the procedure may be repeated indefinitely.

• Can you make a magic triangle? Arrange the digits from 1 through 9 along the sides of a triangle to make the sum of the numbers on each side 23:

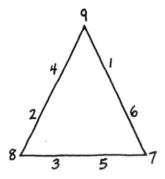

• Place a square within a square so that the diagonals of the small square are parts of the diagonals of the large square (see illustration). Now arrange the digits from 1 through 9 along the diagonals so that the sums of the numbers along the diagonals of the smaller square are equal (12) and the sums of the numbers along the diagonals of the larger square are also equal (23):

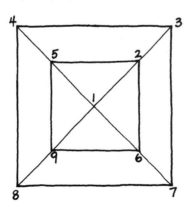

257

• Draw and cut out a figure composed of five squares, as shown. With two straight cuts of a scissors can you cut the figure into three pieces which, fitted together, form a square?

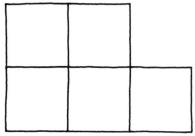

Cut along the dotted lines as shown. Then the three pieces fit together to form the square shown.

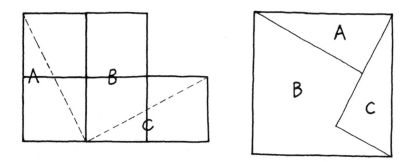

• The four figures illustrated are congruent. Can you fit them together to make a large similar piece?

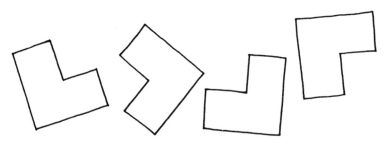

Here's how:

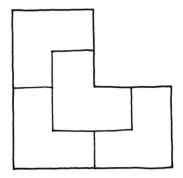

• Can you draw two straight lines on this square to form four congruent quadrilaterals? (No more squares, please.)

So long as *AB* is perpendicular to *CD*, quadrilaterals *E*, *F*, *G*, *H* are congruent no matter how much *CD* tilts from the vertical.

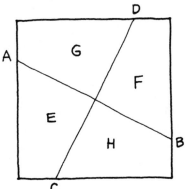

• The house was constructed with ten straight lines. Its front, which is a pentagon, faces right. Can you make it face left by changing the positions of two lines?

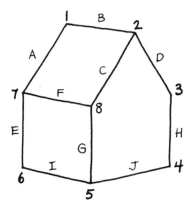

First move line *C* so that it connects points 1 and 8. Then move line *F* so that it connects points 8 and 3:

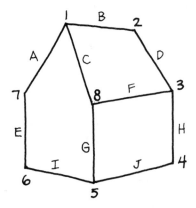

• Can you convert a circle into an equilateral triangle?

Draw the circle with a compass and cut it out with a scissors. To find the center of the circle, fold it in half, open it, and fold it in half at a different place, being careful both times that the rims of the halves are in perfect align-

260

ment. The place where the two creases cross is the center *O* of the circle, as shown.

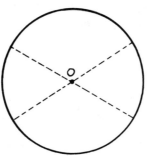

Now fold so that the rim of the circle touches the center, forming a crease with ends *A* and *B*:

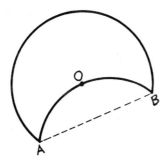

Make a second fold so that the rim touches the center and the crease so formed has one end at *A*. Call the other end *C*:

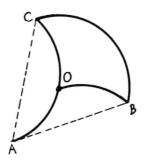

The last fold also makes the rim touch the center and the crease so formed has its ends at B and C. The result is an equilateral triangle:

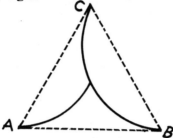

• By creasing and knotting paper you can form a number of polygons — for example, a square.

Begin with two strips of paper of the same width. Fold each through the other to form a loop, as shown.

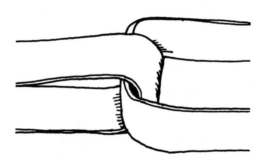

Insert an end of one strip into the loop of the other so that the strips interlock, and pull them tightly together.

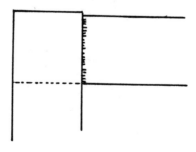

After you cut off the surplus paper or tuck it into the folds, you have a square.

• You can form a regular pentagon (a figure with five equal sides), similarly.

Begin with a long strip of paper of constant width. Loop it into a knot:

Now tie the knot, tighten it, and crease the paper flat:

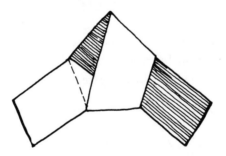

After you snip off or tuck in the surplus and make the creases firm, you have a regular pentagon.

• The regular hexagon (six equal sides) starts with two long strips of paper of exactly equal width. Tie them in a knot:

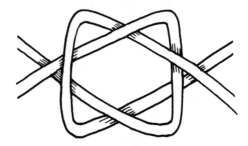

Tighten the knot and crease it flat:

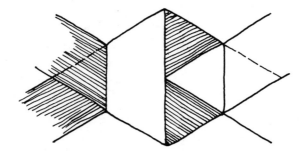

After you trim off the surplus paper and tuck the rest inside the folds, you have a regular hexagon, which can be made quite firm with a little paste.

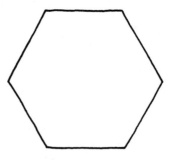

• Can you find in the figure at least one of each of the following geometrical shapes: a square, a rectangle, a parallelogram, a trapezoid, an isosceles triangle, an equilateral triangle, a pentagon, a hexagon, a circle, a semicircle? How about a 30°, 60°, 90°, and 120° angle, and a 60° and a 120° arc?

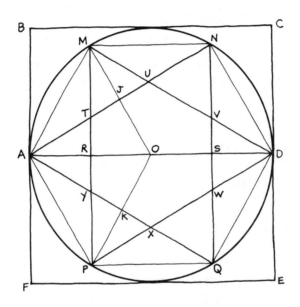

Squares: *BCEF, MNSR, RSQP*. Rectangles: *MNQP, ANDP, MDQA*. Parallelograms: *MNDO, ODQP, AUDX, AMOP*. Trapezoids: *AMND, ANWX, ADQP*.

Isosceles triangles: *PDQ, POD, MOP, MNA, PXQ, NDV, MUN, QWD*. Equilateral triangles: *ANQ, MDP, AMO, AOP, XWQ, YPX, NUV, MTU, TAY*. Pentagons: *AMNQP, AMNWP, AMVWP, AUDQP*, etc. Hexagon: *AMNDQP*. Circle: *A*. Semicircles: *AMD, APD, MNQ, MAQ, NDP, NMP*.

30° angles: *MAN, PAQ, APM, VND, VDN, ADP, MDA,*

MNU, UMN, YPK, KPX, PDQ, WQD, AMR, TMJ, RMO, OMD, PAK, KAO, PQX, QPX, QAO, TAR.

60° angles: *MAO, MOA, PAO, APO, POA, ANQ, NQA, QAN, MDP, DPM, PMD, XWQ, NWD, WQX, QXW, YPX, YXP, PYX, NUV, UVN, VNU, MTU, TUM, UMT, TAY, AYT, YTA, ATR, AQN.*

60° arcs are subtended by *AOM* and *AOP* and cut by chords *MA, MN, ND, DQ, QP,* and *PA*.

90° angles: *MRO, PRO, ARM, ARP, ASN, ASQ, NSD, QSD, MPQ, PQN, AND, QNM, AMD, NMP, AJO, NJO, MJT, MJU, AKO, AKP, PKQ, OKQ, BAD, FAD, MAQ, NAP, BFE, FEC, ECB, APD, ABC.*

120° angles: *MOD, PXQ, AXD, POD, MUN, AUD, MTA, RTJ, AYP, MYQ, NVD, MVQ, AMN, MND, APQ, PAM, DWQ, PWS.*

120° arcs are subtended by *MOD* and *POD* and cut by chords *MD, AN, AQ,* and *PD*.

• Can you place eight dots in the checkerboard squares so that no two dots are on the same line vertically, horizontally, or diagonally across the midpoints of two boxes?

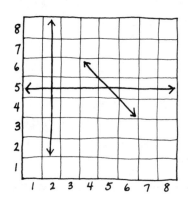

The eight dots should be on squares (1, 6), (2, 3), (3, 7), (4, 2), (5, 8), (6, 5), (7, 1), and (8, 4).

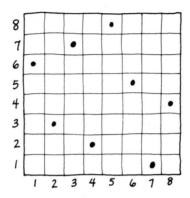

• How do you draw an ellipse?

Stick two pins or tacks into a sheet of paper, a few inches apart. Make a loop of string that goes over them with a little slack. Stick a pencil in the loop, stretch the string tight with the pencil, and sweep around the pins, keeping the string taut. The pencil point will trace out a perfect ellipse.

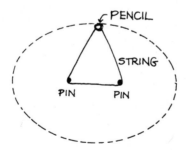

• By drawing two straight lines, can you divide the figure here into four parts, each of the same size and shape?

267

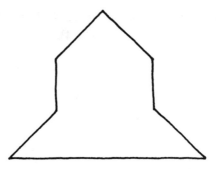

AB and CD divide the figure into four equal trapezoids:

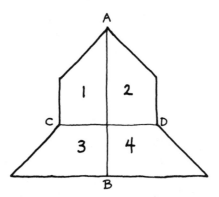

Take the four aces, kings, queens, and jacks of a deck of
playing cards and arrange them in a square, four cards by
four cards, so that neither the same suit nor the same
denomination occurs twice in any four-card row or column:

Col. 1: ace of hearts   Col. 2: king of diamonds
        queen of spades   jack of clubs
        king of clubs   ace of spades
        jack of diamonds   queen of hearts

Col. 3: queen of clubs   Col. 4: jack of spades
        ace of diamonds   king of hearts
        jack of hearts   queen of diamonds
        king of spades   ace of clubs

Here are seven common geometric figures. Under each figure write the first letter of its name. The letters should spell a word. If you are right, the word will show you are

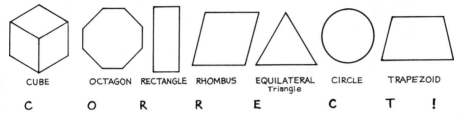

CUBE	OCTAGON	RECTANGLE	RHOMBUS	EQUILATERAL Triangle	CIRCLE	TRAPEZOID
C	O	R	R	E	C	T	!

10

Mirthful Math

HUMOR CAN SERVE as a kind of spice to give a child a taste for learning and to sharpen his appetite for it. Especially when the effort to solve problems involving abstract quantities and spatial relations threatens to produce feelings of frustration or a dislike of mathematics, a bit of pleasantry or nonsense, a quip, or a riddle—anything that will produce a laugh, a giggle, or even a smile—comes as a welcome break.

Any amusing diversion relaxes both body and mind. Interjected at the right moment and used with discretion, jokes can overcome boredom, counteract mental fatigue, and establish cheerful associations with numbers and geometric shapes.

The jests, gags, and wisecracks in this chapter—we might call them "a mirthful mouthful of math"—can be employed as needed to lighten the burden on the mind, brighten the outlook, and refresh the spirit in the midst of learning about figures and their ways.

• Do you say that figures don't lie? Well, they do! Using a pocket calculator, subtract 83 from 400. Now turn the calculator upside down and read what it tells you.

• Why isn't your nose 12 inches long?

Because then it would be a foot.

• When do you have a 10 to 1 chance of catching a train?

When your watch shows 12:50.

• What is more wonderful than a counting dog?

A spelling bee.

• If I were to divide 7654 by 328, what would I get?

The wrong answer.

• What's the difference between 1 yard and 2 yards?

A fence.

• What is the area of a triangle with a base of 10 inches, one side of 6 inches, and one side of 3 inches?

There is no such triangle. Since a straight line is the shortest distance between two points, each side of a triangle must be shorter than the sum of the other two sides. But 10 is greater than 6 + 3.

• Put this question to a friend: "If two men made a bet to see who could eat more nuts, and one ate ninety-nine and the other ate a hundred and won, how many more nuts did the winner eat than the loser?"

The chances are he will think you said "ate a hundred and one" and answer "Two." But your answer is 1.

• What did the adding machine say to the bookkeeper?

"You can count on me."

• What is the right angle from which to approach any problem?

The try angle.

• Little Abe: "Our cat has nine lives."

Little Sam: "That's nothing. We have a frog that croaks every night."

• When do 8 and 3 make more than 11?

When they are written as 83.

• What can you always count on?

Your fingers.

• Put this question to a friend: "Which would you rather have — an old five-dollar bill or a new one?"

If he chooses the new one, you can point out that an old *five*-dollar bill is worth four more dollars than a new *one*.

• Can you prove that $40 \times 2 = 41 \times 2$?

Forty times two equals eighty, and forty-one times two equals eighty-*too*.

• Teacher: How much is 6 and 4?

Student: 11.

Teacher: 6 and 4 is 10.

Student: 6 and 4 can't be 10, because 5 and 5 is 10.

272

• If a barrel weighs 15 pounds, what would you fill it with to make it weigh only 10?

Holes.

• Student: What did I make in arithmetic?

Teacher: Mistakes.

• Take two apples from three apples. How many apples do you have?

Two apples.

• If a doctor gives you three pills and tells you to take one every half hour, how long do they last?

One hour. You take one now, the second a half hour from now, and the third an hour from now.

• Some months have thirty days; some have thirty-one. How many have twenty-eight days?

Twelve.

• Teacher: What's the formula for water?

Student: H, I, J, K, L, M, N, O.

Teacher: That's not the formula I gave you.

Student: But you said it was H to O.

• When is the time on a clock like the whistle on a train?

When it's two to two.

• What digit gets larger when you turn it upside down?

6. Turned upside down it becomes 9.

• What number gives advice to the impatient?

1028, or "one, nought, two, eight."

• Teacher: What's the point of studying mathematics?
Student: The decimal point.

• What is bought by the yard and worn by the foot?
A carpet.

• How many apples were eaten in the Garden of Eden?

Eve *ate,* and Adam *too,* and the Devil *won.* That makes eleven.

• Teacher: How did you find the questions on the math exam?

Student: Easy, but I had trouble with the answers.

• Student: I don't think I deserved 0 in mathematics.

Teacher: Neither do I, but it's the lowest mark I can give you.

• When does a bicycle go as fast as an airplane?

When it is in the airplane.

• Why can't a clock run forever?

Because its hours are numbered.

• If I cut a piece of steak into two parts, what do I get?

Halves.

And four parts?

Quarters.

And a thousand parts?

Hamburger.

• What number is a sign of danger?

4. ("Fore" in golf is a warning to those ahead that a ball is about to be driven in their direction.)

• Wife: Please lend me twenty dollars, dear.

Husband: I'm sorry, but all I have is ten.

Wife: That will be fine. Just give me ten of the twenty I asked you for. Then, as I owe you ten and you owe me ten, we'll call it square.

• What did the snakes say when they were told to "go forth and multiply"?

"We can't. We're adders."

• Student: You flunked me in arithmetic. I can't understand it.

Teacher: That's why I flunked you.

• Visitor: What's your name?

Boy: 6 7/8.

Visitor: What an odd name! Where did your parents get it?

Boy: They pulled it out of a hat.

• Teacher: Tom, if you had two dollars and asked your

father for one more dollar, how many dollars would you have?

Tom: Two dollars.

Teacher: You don't know your arithmetic very well.

Tom: You don't know my father very well.

• Can you define a vacation, using the number 2?

A vacation is 2 weeks that are 2 short, after which you are 2 tired 2 go home and 2 broke not 2.

• What did the fly say when he was swatted as he landed on the adding machine?

"I guess my number was up."

APPENDIX

Symbols Commonly Used in Geometry

$\geqq$ Equal to or greater than

$\leqq$ Equal to or less than

$\cong$ Congruent

$\llcorner$ Right angle

$\times$ Vertical angles

     Alternate interior angles

Corresponding angles

 Isosceles triangle

 Equilateral triangle

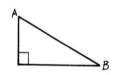 Hypotenuse (*AB*)

 Parallel lines

 Is similar to

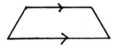 Trapezoid

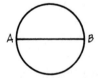 Diameter (*AB*) of a circle

278

 Inscribed angle

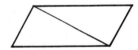 Secant (*AB*) and tangent (*CD*)

 Chord of a circle

 Adjacent angles (*a*, *b*)

 Diagonal

 Complementary angles (*a*, *b*)

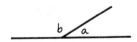 Supplementary angles (*a*, *b*)

Formulas for Calculating the Areas
of Common Geometric Figures

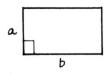
Rectangle $\qquad$ $A = ab$

Square $\qquad$ $A = s^2$

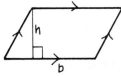
Parallelogram $\qquad$ $A = bh$

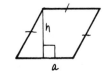
Rhombus $\qquad$ $A = ah$

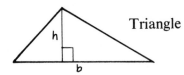
Triangle $\qquad$ $A = \dfrac{1}{2}\,bh$

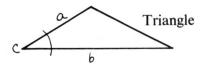
Triangle $\qquad$ $A = \dfrac{ab\,\sin C}{2}$

Equilateral triangle $A = \dfrac{s^2\sqrt{3}}{4}$

 Trapezoid $A = \dfrac{h(a+b)}{2}$

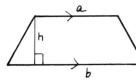 Circle $A = \pi r^2 (\pi$ = approx. 3.1416)

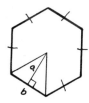 Regular pentagon (where p is the perimeter) $A = \dfrac{ap}{2} = \dfrac{5ab}{2}$

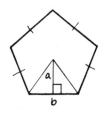 Regular hexagon (where p is the perimeter) $A = \dfrac{ap}{2} = 3ab = \dfrac{3b^2\sqrt{3}}{2}$ (six equilateral triangles)

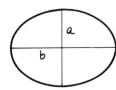 Ellipse (where a and b are the axes) $A = \pi ab$

Formulas for Calculating the Volumes of Common Geometric Solids

Cube
$V = s^3$

Pyramid
(where
h = altitude
and B = area
of base, which
may be any
plane figure)
$V = \dfrac{Bh}{3}$

Rectangular
prism
$V = lwh$

Right circular
cone
$V = \dfrac{\pi r^2 h}{3}$

Right circular
cylinder
$V = \pi r^2 h$

Sphere
$V = \dfrac{4\pi r^3}{3}$

Table of Metric Equivalents

Linear Measure

1 inch = 2.54 centimeters (cm) = 25.4 millimeters (mm)
1 foot = 0.3048 meter (m) = 30.48 centimeters = 304.8 millimeters
1 yard = 0.9144 meter = 91.44 centimeters
1 mile = 1.6093 kilometers (km)
50 miles per hour = 80.46 kilometers per hour
1 millimeter = 0.001 meter = 0.1 centimeter = 0.03937 inch
1 centimeter = 0.01 meter = 10 millimeters = 0.3937 inch
1 meter = 100 centimeters = 1000 millimeters = 39.37 inches
1 kilometer = 1000 meters = 0.62137 mile

Square Measure

1 square inch = 6.4516 square centimeters
1 square foot = 0.0929 square meter
1 square yard = 0.8361 square meter
1 acre = 0.4047 hectare
1 square mile = 258.9 hectares
1 square centimeter = 0.155 square inch
1 square meter = 1.196 square yards
1 hectare = 10,000 square meters = 2.471 acres
1 square kilometer = 0.3861 square mile

Cubic Measure

1 cubic inch = 16.387 cubic centimeters
1 cubic foot = 0.0283 cubic meter
1 cubic yard = 0.7646 cubic meter
1 cubic centimeter = 0.061 cubic inch

Liquid Measure

1 pint = 0.473 liter
1 quart = 0.946 liter
1 gallon = 3.785 liters
1 liter = 0.264 gallon = 1.0567 quarts

Weight Measure

1 ounce avoirdupois = 28.349 grams (gm)
1 pound avoirdupois = 0.4536 kilogram (kg) = 453.6 grams
1 short ton = 0.9072 metric ton
1 gram = 0.035 ounce avoirdupois
1 kilogram = 2.2046 pounds avoirdupois = 35.273 ounces
  avoirdupois
1 metric ton = 1.102 short tons = 2204.623 pounds
  avoirdupois

In the metric system the prefix *deka-* signifies a multiple
of 10: 1 dekameter = 10 meters; 1 dekaliter = 10 liters;
1 dekagram = 10 grams.

The prefix *hecto-* signifies a multiple of 100: 1 hectometer =
100 meters; 1 hectoliter = 100 liters; 1 hectogram = 100
grams.

The prefix *kilo-* signifies a multiple of 1000: 1 kilometer =
1000 meters; 1 kiloliter = 1000 liters; 1 kilogram = 1000
grams.

The prefix *deci-* signifies a multiple of one-tenth (0.1):
1 decimeter = 0.1 meter; 1 deciliter = 0.1 liter; 1 deci-
gram = 0.1 gram.

The prefix *centi-* signifies a multiple of one-hundredth (0.01): 1 centimeter = 0.01 meter; 1 centiliter = 0.01 liter; 1 centigram = 0.01 gram.

The prefix *milli-* signifies a multiple of one-thousandth (0.001): 1 millimeter = 0.001 meter; 1 milliliter = 0.001 liter; 1 milligram = 0.001 gram.

Temperature

To convert degrees centigrade (also called degrees Celsius, after the Swedish astronomer who invented the centigrade thermometer) to degrees Fahrenheit, multiply by 9/5 and add 32: 0°C = 32°F; 100°C = 212°F.

To convert degrees Fahrenheit to degrees centigrade, subtract 32 and then multiply by 5/9: (98.6°F − 32) × 5/9 = 37°C.

Squares and Square Roots

	Square	Square Root		Square	Square Root		Square	Square Root
1	1	1.000	**51**	2601	7.141	**101**	10,201	10.050
2	4	1.414	**52**	2704	7.211	**102**	10,404	10.100
3	9	1.732	**53**	2809	7.280	**103**	10,609	10.149
4	16	2.000	**54**	2916	7.348	**104**	10,816	10.198
5	25	2.236	**55**	3025	7.416	**105**	11,025	10.247
6	36	2.449	**56**	3136	7.483	**106**	11,236	10.296
7	49	2.646	**57**	3249	7.550	**107**	11,449	10.344
8	64	2.828	**58**	3364	7.616	**108**	11,664	10.392
9	81	3.000	**59**	3481	7.681	**109**	11,881	10.440
10	100	3.162	**60**	3600	7.746	**110**	12,100	10.488
11	121	3.317	**61**	3721	7.810	**111**	12,321	10.536
12	144	3.464	**62**	3844	7.874	**112**	12,544	10.583
13	169	3.606	**63**	3969	7.937	**113**	12,769	10.630
14	196	3.742	**64**	4096	8.000	**114**	12,996	10.677
15	225	3.873	**65**	4225	8.062	**115**	13,225	10.724
16	256	4.000	**66**	4356	8.124	**116**	13,456	10.770
17	289	4.123	**67**	4489	8.185	**117**	13,689	10.817
18	324	4.243	**68**	4624	8.246	**118**	13,924	10.863
19	361	4.359	**69**	4761	8.307	**119**	14,161	10.909
20	400	4.472	**70**	4900	8.367	**120**	14,400	10.954
21	441	4.583	**71**	5041	8.426	**121**	14,641	11.000
22	484	4.690	**72**	5184	8.485	**122**	14,884	11.045
23	529	4.796	**73**	5329	8.544	**123**	15,129	11.091
24	576	4.899	**74**	5476	8.602	**124**	15,376	11.136
25	625	5.000	**75**	5625	8.660	**125**	15,625	11.180
26	676	5.099	**76**	5776	8.718	**126**	15,876	11.225
27	729	5.196	**77**	5929	8.775	**127**	16,129	11.269
28	784	5.292	**78**	6084	8.832	**128**	16,384	11.314
29	841	5.385	**79**	6241	8.888	**129**	16,641	11.358
30	900	5.477	**80**	6400	8.944	**130**	16,900	11.402
31	961	5.568	**81**	6561	9.000	**131**	17,161	11.446
32	1024	5.657	**82**	6724	9.055	**132**	17,424	11.489
33	1089	5.745	**83**	6889	9.110	**133**	17,689	11.533
34	1156	5.831	**84**	7056	9.165	**134**	17,956	11.576
35	1225	5.916	**85**	7225	9.220	**135**	18,225	11.619
36	1296	6.000	**86**	7396	9.274	**136**	18,496	11.662
37	1369	6.083	**87**	7569	9.327	**137**	18,769	11.705
38	1444	6.164	**88**	7744	9.381	**138**	19,044	11.747
39	1521	6.245	**89**	7921	9.434	**139**	19,321	11.790
40	1600	6.325	**90**	8100	9.487	**140**	19,600	11.832
41	1681	6.403	**91**	8281	9.539	**141**	19,881	11.874
42	1764	6.481	**92**	8464	9.592	**142**	20,164	11.916
43	1849	6.557	**93**	8649	9.644	**143**	20,449	11.958
44	1936	6.633	**94**	8836	9.695	**144**	20,736	12.000
45	2025	6.708	**95**	9025	9.747	**145**	21,025	12.042
46	2116	6.782	**96**	9216	9.798	**146**	21,316	12.083
47	2209	6.856	**97**	9409	9.849	**147**	21,609	12.124
48	2304	6.928	**98**	9604	9.899	**148**	21,904	12.166
49	2401	7.000	**99**	9801	9.950	**149**	22,201	22.207
50	2500	7.071	**100**	10000	10.000	**150**	22,500	12.247

Values of the Trigonometric Functions

Angle	Sine	Cosine	Tangent	Angle	Sine	Cosine	Tangent
1°	0.0175	0.9998	0.0175	46°	0.7193	0.6947	1.0355
2°	0.0349	0.9994	0.0349	47°	0.7314	0.6820	1.0724
3°	0.0523	0.9986	0.0524	48°	0.7431	0.6691	1.1106
4°	0.0698	0.9976	0.0699	49°	0.7547	0.6561	1.1504
5°	0.0872	0.9962	0.0875	50°	0.7660	0.6428	1.1918
6°	0.1045	0.9945	0.1051	51°	0.7771	0.6293	1.2349
7°	0.1219	0.9925	0.1228	52°	0.7880	0.6157	1.2799
8°	0.1392	0.9903	0.1405	53°	0.7986	0.6018	1.3270
9°	0.1564	0.9877	0.1584	54°	0.8090	0.5878	1.3764
10°	0.1736	0.9848	0.1763	55°	0.8192	0.5736	1.4281
11°	0.1908	0.9816	0.1944	56°	0.8290	0.5592	1.4826
12°	0.2079	0.9781	0.2126	57°	0.8387	0.5446	1.5399
13°	0.2250	0.9744	0.2309	58°	0.8480	0.5299	1.6003
14°	0.2419	0.9703	0.2493	59°	0.8572	0.5150	1.6643
15°	0.2588	0.9659	0.2679	60°	0.8660	0.5000	1.7321
16°	0.2756	0.9613	0.2867	61°	0.8746	0.4848	1.8040
17°	0.2924	0.9563	0.3057	62°	0.8829	0.4695	1.8807
18°	0.3090	0.9511	0.3249	63°	0.8910	0.4540	1.9626
19°	0.3256	0.9455	0.3443	64°	0.8988	0.4384	2.0503
20°	0.3420	0.9397	0.3640	65°	0.9063	0.4226	2.1445
21°	0.3584	0.9336	0.3839	66°	0.9135	0.4067	2.2460
22°	0.3746	0.9272	0.4040	67°	0.9205	0.3907	2.3559
23°	0.3907	0.9205	0.4245	68°	0.9272	0.3746	2.4751
24°	0.4067	0.9135	0.4452	69°	0.9336	0.3584	2.6051
25°	0.4226	0.9063	0.4663	70°	0.9397	0.3420	2.7475
26°	0.4384	0.8988	0.4877	71°	0.9455	0.3256	2.9042
27°	0.4540	0.8910	0.5095	72°	0.9511	0.3090	3.0777
28°	0.4695	0.8829	0.5317	73°	0.9563	0.2924	3.2709
29°	0.4848	0.8746	0.5543	74°	0.9613	0.2756	3.4874
30°	0.5000	0.8660	0.5774	75°	0.9659	0.2588	3.7321
31°	0.5150	0.8572	0.6009	76°	0.9703	0.2419	4.0108
32°	0.5299	0.8480	0.6249	77°	0.9744	0.2250	4.3315
33°	0.5446	0.8387	0.6494	78°	0.9781	0.2079	4.7046
34°	0.5592	0.8290	0.6745	79°	0.9816	0.1908	5.1446
35°	0.5736	0.8192	0.7002	80°	0.9848	0.1736	5.6713
36°	0.5878	0.8090	0.7265	81°	0.9877	0.1564	6.3138
37°	0.6018	0.7986	0.7536	82°	0.9903	0.1392	7.1154
38°	0.6157	0.7880	0.7813	83°	0.9925	0.1219	8.1443
39°	0.6293	0.7771	0.8098	84°	0.9945	0.1045	9.5144
40°	0.6428	0.7660	0.8391	85°	0.9962	0.0872	11.4301
41°	0.6561	0.7547	0.8693	86°	0.9976	0.0698	14.3007
42°	0.6691	0.7431	0.9004	87°	0.9986	0.0523	19.0811
43°	0.6820	0.7314	0.9325	88°	0.9994	0.0349	28.6363
44°	0.6947	0.7193	0.9657	89°	0.9998	0.0175	57.2900
45°	0.7071	0.7071	1.0000	90°	1.0000	0.0000	

Index of Skills Taught

EXTRACTING SQUARE ROOTS

FACTORIAL OPERATIONS

FACTORING

FRACTION OPERATIONS

MULTIPLICATION

USING POSITIVE AND NEGATIVE NUMBERS

USING ROMAN NUMERALS

Index of Games